World Out of BALANCE
Our Polluted Planet

World Out of
BALANCE
Our Polluted Planet
Jon Erickson

TAB | TAB BOOKS
Blue Ridge Summit, PA

FIRST EDITION
FIRST PRINTING

© 1992 by **TAB Books**.
TAB Books is a division of McGraw-Hill, Inc.

Library of Congress Cataloging-in-Publication Data

Erickson, Jon, 1948-
    World out of balance : our polluted planet / by Jon Erickson.
        p.   cm.
    Includes index.
    ISBN 0-8306-2823-1   ISBN 0-8306-2804-5 (pbk.)
    1. Pollution. 2. Man—Influence on nature. I. Title.
    TD174.E75 1992
    363.73—dc20   91-31418
                                                  CIP

TAB Books offers software for sale. For information and a catalog, please contact
TAB Software Department, Blue Ridge Summit, PA 17294-0850.

Acquisitions Editor: Roland S. Phelps
Book Editor: Andrew Yoder
Production: Katherine G. Brown
Series Design: Jaclyn J. Boone
Cover Design: Lori E. Schlosser

Contents

Acknowledgments

The author thanks the following organizations for supplying photographs for this book: the National Aeronautics and Space Administration, the National Oceanic and Atmospheric Administration, the National Optical Astronomy Observatories, the National Park Service, the U.S. Air Force, the U.S. Army Corps of Engineers, the U.S. Coast Guard, the U.S. Department of Agriculture Soil Conservation Service, the U.S. Department of Energy, the U.S. Forest Service, the U.S. Maritime Administration, the U.S. Navy, and the U.S. Geological Survey.

Introduction

Pollution is a problem of global proportions that requires global solutions. Global pollution is so pervasive that more conservation can only attack part of the problem, not solve it. Major changes are occurring and our species is mostly to blame for the environmental disturbances that beset our planet. We are conducting a global experiment by changing the face of the entire world. We are destroying the rain forests and pumping our pollutants into the air and water, thus unfavorably altering the composition of the biosphere and upsetting the Earth's heat budget.

Global pollution is rising so rapidly that the natural forces needed to return everything back into balance are being overwhelmed. Increasingly, air contaminants are emitted into the atmosphere and dangerous toxic substances are released into the ocean. We are rapidly running out of space to put our wastes. Highly destructive acid rain is increasing. The protective ozone layer is eroding. When it goes, we go.

Millions of species of plants and animals are already dying as a result of human encroachment on their environment. By the middle of the next century, as many as half the species living on Earth today could become extinct—never to be seen again. It would be a great tragedy if through our neglect of the environment and wanton destruction of life, we would force vast numbers of species into extinction and return the planet to a condition of low diversity.

If we do not curb our insatiable appetite for fossil fuels and stop destroying the forests and their wildlife, by the middle of the next century, the world could be hotter than it has been in the past million years. Some of our man-made greenhouse gases are extremely toxic and carcinogenic; others are destroying the ozone layer.

World populations are growing so explosively and modifying the environment so extensively that we are inflicting a global impact of unprecedented dimensions. Rapid population growth has already stretched the resources of the world. The prospect of future increases raises serious doubts as to whether the planet can continue to support people's growing needs. Up to a tenfold increase in world economic activity over the next 50 years would be required to keep pace with basic human requirements. The biosphere cannot possibly tolerate this situation without irreversible damage.

Already, dramatic changes are occurring as a result of the improper use of land and water resources, large-scale extraction and combustion of fossil fuels, and widespread use of chemicals in industry and agriculture. These activities could permanently alter the balance of nature. The delicate and complicated interdependence that organisms have on each other and on their environment is not yet fully understood. However, it is becoming more apparent that if we continue to upset nature's balance through our wanton negligence and waste, we will be left with a world that would not be to our liking.

About the Author

Jon Erickson has written several earth science books for TAB, including the highly acclaimed *Discovering Earth Science Series*. He holds degrees in physical science and natural science, and has worked as a U.S. Navy radar technician, as a geologist for major oil and mining companies, and as an engineer for an aerospace company and an electronic instruments company. He presently lives in western Colorado, where he works as an independent geologist, consulting technical editor, and writer.

OTHER TAB BOOKS BY THE AUTHOR:

Volcanoes and Earthquakes
Violent Storms
Mysterious Oceans
The Living Earth
Exploring Earth From Space
Ice Ages
Greenhouse Earth
Target Earth!
Dying Planet

1

The Polluted Planet

THE Earth has been exceptionally hospitable for mankind, providing abundant water and other necessities at his disposal. Our planet also happens to lie in the narrow temperature range where water remains a liquid. The Earth is encircled by a thin gaseous outer layer, called the *atmosphere*, which has a large amount of *oxygen*, a gas that is almost entirely missing on other planets. If Earth was as cold and dry as Mars or as hot and steamy as Venus, it is highly doubtful that life would have formed.

No other species has affected this planet during its long history than man. Increasing human populations are inflicting an environmental disaster of unprecedented dimensions. Global pollution is rising so rapidly that natural forces needed to bring back the balance are being overwhelmed. The human race will soon feel the repercussions of our destructive practices, along with the rest of the living world. If we destroy Earth, as well as ourselves, no one will be left to remark on how hostile the planet has become.

THE PREHISTORIC AGE

The Earth was born during a time of an unusual stellar formation. Probably less than 10 percent of all stars in the galaxy are single, main-sequence stars, and fewer yet have a system of orbiting planets. Our Solar System consists of nine planets and their satellites (FIG. 1-1). The inner terrestrial planets have similar compositions, with the notable exception of Earth, which possesses an ocean of water and a breathable atmosphere. It is also the only terrestrial planet with a rather large moon, the biggest in relation to its mother planet, which still defies explanation.

NASA

Fig. 1-1. The Solar System, showing the Earth, lunar surface, Sun, Mercury, Venus, Mars, Jupiter, and Saturn.

The position of Earth in the Solar System also has a special significance. It is not so close to the Sun that conditions would be extremely hot. Nor is it so far away that conditions would be extremely cold. The Earth's orbit is nearly circular, varying by no more than about 3 percent. A widely elliptical orbit would swing the planet close by the Sun in one season (searing it) and toss it far away in the opposite season (freezing it), a situation that would not be very conducive to life.

The atmosphere and ocean evolved during a period of crustal formation and volcanic outgassing. In addition, multitudes of giant meteorites slammed into Earth, adding their own special ingredients to the boiling caldron. The Moon and other bodies in the Solar System still show numerous scars from this massive invasion (FIG. 1-2). However, because Earth developed an active weathering system early on, its meteorite craters have long since been erased. Raging storms brought deluge after deluge and unimaginable electrical displays.

Out of this chaos came life. An unusual set of circumstances that cannot be duplicated no matter how hard scientists try brought organic molecules together in the proper sequence, catalyzed by lightning bolts, and the molecules were washed into the seas. It is amazing how life managed to survive in all this turmoil. Heavy meteorite bombardments, massive volcanic eruptions, and large accumulations of toxic substances in the

Fig. 1-2. The Earth's moon, showing numerous craters from a massive meteorite bombardment four billion years ago.

ocean must have created a pollution problem of immense proportions. Yet, life struggled through it all.

It is fortunate that the early Earth did not have significant levels of oxygen in the atmosphere and ocean. If it did, life could not have been created because oxygen would have poisoned all primitive lifeforms (TABLE 1-1). If the oxygen level had continued to rise without some means of removing it or if organisms had not evolved mechanisms for coping with it, the Earth might have witnessed massive pollution and the death of all life. If none of the oxygen in our present atmosphere had been removed by respiration and decay, its level would double in 10,000 years and the Earth would be incinerated.

The absence of atmospheric oxygen also left the earth exposed to high levels of solar ultraviolet (UV) radiation as a result of the lack of an ozone layer. This was not a major problem at that time, however, because simple organisms, called *prokaryotes*, could withstand significantly higher levels of UV than more complex organisms, called *eukaryotes*. The proliferation of oxygen-producing plankton between 1.5 and 2.5 billion

TABLE 1-1. Development of Life on Earth.		
EVOLUTION	ORIGIN (MILLION YEARS)	ATMOSPHERE
Origin of Earth	4600	Hydrogen, helium
Origin of life	3800	Nitrogen, methane, carbon dioxide
Photosynthesis	2300	Nitrogen, carbon dioxide, oxygen
Eukaryotic cells	1400	Nitrogen, carbon dioxide, oxygen
Sexual reproduction	1100	Nitrogen, oxygen, carbon dioxide
Metazoans	700	Nitrogen, oxygen
Land plants	400	Nitrogen, oxygen
Land animals	350	Nitrogen, oxygen
Mammals	200	Nitrogen, oxygen
Man	2	Nitrogen, oxygen

years ago was probably related to the generation of an ozone shield as the amount of atmospheric oxygen began to rise and the level of biologically harmful UV began to fall.

Only when conditions were right and oxygen began to make up a significant portion of the atmosphere did life develop into complex lifeforms. After life had been around for nearly 80 percent of Earth history (FIG. 1-3), the first shelled invertebrates appeared. These were followed by the first vertebrates, and the fishes claimed dominion over the seas. Scientists believe that some fishes developed lungs and leglike appendages and crawled out of the sea to make a new home on dry land. They were greeted by a garden paradise, replete with luxuriant plant growth.

These primitive amphibians dictated the basic body form that almost all land animals would assume from then on. This body form included bilateral symmetry, four legs each with five toes on each foot, a head with stereo vision and hearing, and a tail. The amphibians were the first permanent land dwellers, and they proliferated during a time when much of Earth was covered with swamps. Because the amphibians were restricted to a life in a semiaquatic environment, they gave way to the reptiles when the swamps dried out. The reptiles were brilliantly successful because (unlike the amphibians) they did not have to spend their lives in the water, which allowed them to venture over practically the entire planet.

The reptiles' great success led to the development of the dinosaurs (FIG. 1-4), which took these survival characteristics to great extremes. The climate and prodigious vegetation growth allowed some giant species to grow as tall as multistory buildings. Unfortu-

Fig. 1-3. The geologic time spiral, which depicts the development of life throughout Earth history.

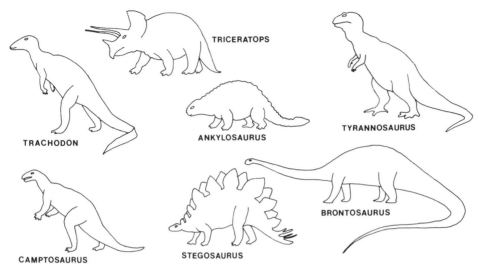

Fig. 1-4. The major Cretaceous dinosaur taxa.

nately, some reason (such as a changing environment) made them all unfit to survive, and suddenly the dinosaurs disappeared after ruling for 140 million years.

Periods of mass extinction are not unusual in the history of Earth (TABLE 1-2), especially since the development of more complex species, which were adapted to a narrow spectrum of living conditions. The worst extinction event in Earth history occurred at the end of the Paleozoic era (240 million years ago), just prior to the dinosaurs. It claimed over 95 percent of all species, leaving the world almost as devoid of life as it was at the beginning of the era, 330 million years earlier (FIG. 1-5).

When the dinosaurs left the stage, the mammals were poised to take over the world. Perhaps the mammal's claim to fame was a greater intelligence, better adaptability, and more complex social behavior. When taken to extremes, these conditions led to the evolution of a most unusual creature. At first, it apparently shared many of the same characteristics as its ancient apelike ancestors. Gradually, however, after being forced to live under more challenging conditions alongside ferocious predators, this two-legged, nearly naked animal became a fierce competitor, able to utilize its environment more capably than any other species that has ever lived.

Scientists believe that humans evolved in Africa from a common ancestor with the chimpanzees 5 to 7 million years ago; today, the two species share 99 percent of the same genes. Life on the savannah, where our ancient ancestors roamed, was much harsher than life in the forests, where the chimpanzees lived. Therefore, in order to survive, the early humans had to develop into intelligent, upright-walking species rather quickly, whereas the chimpanzees, whose environment was less challenging, are much the same today as they were millions of years ago.

THE STONE AGE

Modern humans, called the *Cro-Magnon*, originated in Africa as early as 100,000 years ago. For unknown reasons they did not penetrate further into the Old World for

TABLE 1-2. Radiation and Extinction for Major Organisms.

ORGANISM	RADIATION	EXTINCTION
Marine invertebrates	Lower Paleozoic	Permian
Foraminiferans	Silurian	Permian & Triassic
Graptolites	Ordovician	Silurian & Devonian
Brachiopods	Ordovician	Devonian & Carboniferous
Nautiloids	Ordovician	Mississippian
Ammonoids	Devonian	Upper Cretaceous
Trilobites	Cambrian	Carboniferous & Permian
Crinoids	Ordovician	Upper Permian
Fishes	Devonian	Pennslvanian
Land plants	Devonian	Permian
Insects	Upper Paleozoic	
Amphibians	Pennsylvanian	Permian-Triassic
Reptiles	Permian	Upper Cretaceous
Mammals	Paleocene	Pleistocene

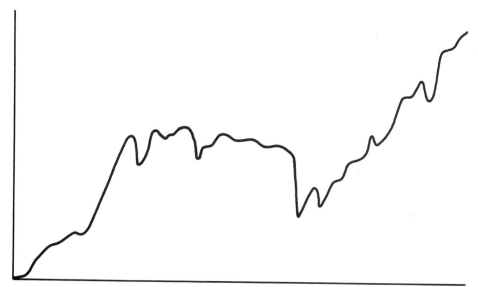

Fig. 1-5. The extinction of species over the last 600 million years. The large dip is in response to the great Permian extinction 240 million years ago.

another 50,000 years. The Cro-Magnon are classified as *Homo sapiens*, the same as us. They were named after the Cro-Magnon cave near Les Eyzies, France, where the first fossil bones were found in 1868. Between 45,000 and 35,000 years ago, during the last ice age, the Cro-Magnon advanced into Europe and Asia. The European geography at this time was mostly grassland and tundra, where plentiful reindeer, woolly mammoths, and other large species roamed. The Cro-Magnon hunted these animals with great skill and carved their carcasses with finely honed blades.

They developed advanced tools, such as stone axes (FIG. 1-6) to cut down trees for fire wood. Their deadly spears, often engraved with animals, became effective weapons

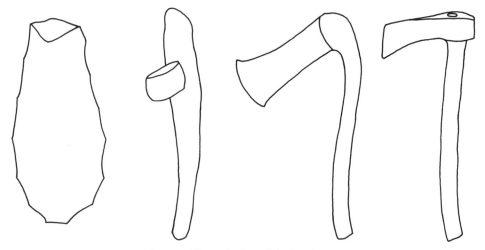

Fig. 1-6. The evolution of the hand ax.

for hunting big game. The Cro-Magnon might have even hunted their stocky cousins, the *Neanderthals*, which would account for their sudden departure approximately 35,000 years ago—about the same time Cro-Magnon came on the scene.

Music played an important role in Cro-Magnon culture during the upper Paleolithic period (from 35,000 to 10,000 years ago). The first-known musical instrument is a bone flute discovered in France that is about 30,000 years old. By about 23,000 years ago, bone sewing needles appeared in southwestern France. The needles were used to tailor cold-weather clothing for coping with the Ice-Age climate.

One of the most important human developments was the trading of goods, especially body ornaments, such as seashells for making necklaces. Jewelry was also made from stone and ivory beads and teeth of dangerous animals, such as bears and lions. Even at this early age, humans developed class distinctions by adorning their bodies with jewelry and other symbols of authority.

Another form of human expression was cave paintings. Some paintings in the American Southwest are believed to have an astronomical purpose. The walls of one cave in the French Pyrenees, contains over 200 human handprints from about 26,000 years ago, mostly with missing fingers. The fingers might have been destroyed by disease or infection, or hacked off in some sort of religious ritual. The best-known art objects are the so-called Venus figures that are approximately 26,000 years old and apparently were small, detailed sculptures of fertility symbols. The later Ice-Age people tended to make elaborate and beautiful cave paintings of animals they highly respected (FIG. 1-7).

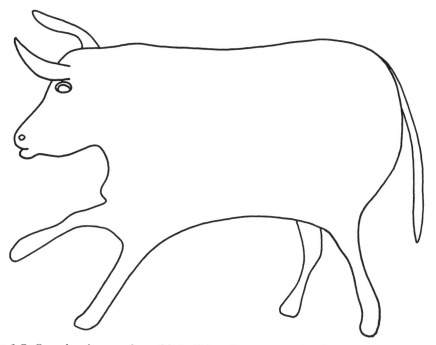

Fig. 1-7. Cave drawings, such as this buffalo, often represented animals that early humans highly regarded.

The Cro-Magnon people became prolific and widespread throughout Europe and Asia during the Ice Age, despite the advancing ice sheets. They might have been attracted to these desolate, frozen lands by a rich stock of large animals that were adapted to the cold. These Ice-Age peoples probably lived much like present-day Eskimos and Lapps, fishing the rivers and hunting reindeer and other large animals. As a result of a lack of wood in the cold tundra to build their homes, Ice-Age hunters of the central Russian plain made shelters with mammoth bones and tusks covered with hides and they burned animal fat for lighting and heating. It is suspected that Ice-Age man over-hunted the mammoth and other large animals, and drove them into extinction (FIG. 1-8).

Ice-Age peoples populated the Americas as early as 32,000 years ago by traveling over a land bridge between Siberia and Alaska across the Bering Strait. The continental ice sheets locked up large quantities of water, which substantially lowered the level of the sea and exposed parts of the shallow seafloor to the surface. From North America, these early Americans crossed over the Panama isthmus into South America and roamed as far south as southern Chile. Cave paintings in Brazil suggest that cave art developed in the Americas at almost the same time it did in Europe and Africa.

Fig. 1-8. The woolly mammoth, like many large mammals, became extinct at the end of the last ice age (about 11,000 years ago) at the same time that humans migrated into their habitats.

THE AGRICULTURAL AGE

When the continental glaciers began retreating about 15,000 years ago, the eastern shores of the Mediterranean witnessed one of the most momentous events in man's history. While hunting deer and wildebeest and gathering food along the North African coast, primitive peoples stumbled on the fertile Levant region that borders on the eastern Mediterranean Sea. They also discovered an area known as the *Fertile Crescent*, that stretches from the southeast coast of the Mediterranean to the Persian Gulf.

Abundant stands of wild wheat and barley grew in thickets on the uplands and people gathered the wild plants and used primitive stone grinders to process the cereals. The stability of this food supply encouraged people to build permanent settlements. They devised tools to harvest the crop and invented pottery to store and cook their food. They herded gazelle, rather than hunt them, which created a system of animal husbandry began.

The Neolithic Age (New Stone Age), which began approximately 10,000 years ago, was the beginning of the food-producing revolution. It occurred virtually simultaneously in the Old World and a few thousand years later in the New World (FIG. 1-9). Man slowly learned the best combinations of resources at hand. Instead of merely gathering wild foods, early farmers collected plants and attempted to control and nurture them. Sheep, goats, pigs, and cattle were domesticated. When resources were exhausted, as they often were, the community totally collapsed, and people were forced to move elsewhere.

The first Neolithic industry was the manufacture of pottery and plaster around 7000 B.C. Plaster was probably invented as early as 12,000 B.C.—long before pottery. Plaster artifacts discovered in the Near East include flooring material, containers, sculptures, and ornamental beads. With the introduction of agriculture, more durable storage vessels for agricultural goods were needed, which led to the invention of pottery.

Megolithic monuments, such as Stonehenge in southern England and the great statues of Easter Island in the Pacific, are among the most dramatic remains of prehis-

Fig. 1-9. Anasazi ruins at Mesa Verde National Park, Colorado. Several hundred years ago, the Anasazi Indians farmed the mesa until a long period of drought apparently drove them away.

toric culture around the world (FIG. 1-10). It appears that many of the hundreds of various circles of tall monoliths found in Europe were for astronomical purposes, such as the telling of the seasons. The oldest monuments date to about 4000 B.C. and are often composed of exotic rock that had been hauled in from great distances away. Such laborious activity might have involved the worship of stones.

Humans discovered fire long before the last Ice Age. However, it is not clear whether they could light a fire by this time; they probably used fires that were set naturally by lightning strikes. For thousands of years, humans did little with fire, except to cook meals, keep warm, and hunt game by setting brush fires to frighten animals into traps or to chase them off cliffs.

About 6000 years ago, people began to use fire to bake pottery and to forge bronze for tools and weapons. Copper and tin were mined extensively on the island of Cyprus by the early Greeks. Mining tools from before 2500 B.C. were found in the underground workings. The rich ore provided some of the first bronze for the earliest Greek sculptures. The Bronze Age subsequently gave way to the Iron Age at about 1000 B.C. All this time, man's use of fire was still rudimentary until the introduction of the steam engine in the late 1700s. The steam engine mated coal and iron and ushered in the Industrial Revolution.

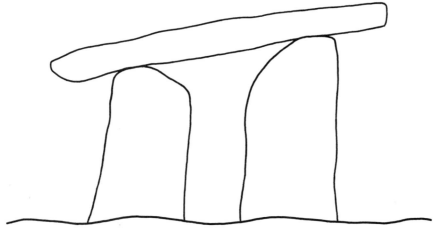

Fig. 1-10. Megalithic monuments that are scattered throughout Europe might have had an astronomical function.

THE INDUSTRIAL AGE

The Industrial Revolution was made through the exploitation of coal, steam, and iron. Before industrialization, the small fires of civilization had virtually no effect on the environment. The fuel used was mostly wood, but as populations continued to grow as a direct result of industrializing, the forests of Europe rapidly disappeared until the timely discovery of coal. Actually, coal had been known for quite some time. Early peoples probably discovered it when lightning ignited outcrops of coal. For ages, American Indians used coal for their cooking fires.

In Europe, coal fueled the engines of the early industrial age. It fed the furnaces of the factories and foundries, which were often located near the coal mines to minimize transportation. Work in the mines was often back-breaking and dangerous and the wages were poor. In America, families settled in company mining towns, where the prices for rent and supplies were often higher than the miners' wages, which forced many into debt.

By the middle of the 18th century, the Industrial Revolution spread throughout Britain and Europe. However, it did not make its debut in the United States for almost another century, because the economy was mostly agrarian. Even at the beginning of the 20th century, 50 percent of the American population still worked on farms. However, when American industrialists finally began to practice their trade, they did so in a big way. New inventions proliferated, including the steam locomotive and the steamship, whose tracks and sea lanes spread throughout the entire world to feed the hungry giant of industry.

Great industries gave rise to huge industrial cities, where coal smoke belched from a profusion of chimneys. Although the public generally benefited from the prosperity promised by industrialization, people also suffered from serious health problems. Smoke was so heavy that people died in unusually large numbers—mainly from lung ailments. Ash and soot covered most everything, blackening buildings, trees, and other objects. Smoke from the numerous factories stung the eyes and the sulfurous stench of burning coal was everywhere. Some days the smoke was so thick that people could not even see the Sun.

The lure of the big cities and better-paying jobs started a wave of migration from the farm to the factory. By the middle 19th century, the populations of major industrialized countries were becoming more urban than rural.

2

The Critical Cycles

MANY aspects of life on Earth are governed by cycles. The continuous evaporation of seawater and the precipitation of rain and snow on the continents is one of nature's most important cycles. It provides much needed water for life on the surface and it cleanses the environment. The circulation of carbon is required to maintain living conditions on the planet. Major changes in the carbon cycle determine whether we live in a hot house or in an ice house. The recycling of nitrogen in the biosphere is also fundamental for the support of life. Nature operates in a delicate balance and any interference by man could upset it, making living conditions difficult for all living things.

THE WATER CYCLE

The movement of water on the Earth is one of nature's most important cycles. Without the flow of water from the ocean, over the land, and back to the sea, no life would exist. The oceans cover about 75 percent of the planet's surface to an average depth of over two miles, amounting to nearly one-quarter of a billion cubic miles of water. Every day, one trillion tons of water rain down on the planet, but most of it falls directly back into the sea.

The average time required for water to travel from the ocean to the atmosphere, across the land, and back to the sea again is about 10 days. The journey requires only a few hours in the tropical coastal areas, but it might be as long as 10,000 years in the polar regions. This circulation of water onto the land and back into the sea is the *hydrologic cycle*, better known as the *water cycle* (FIG. 2-1).

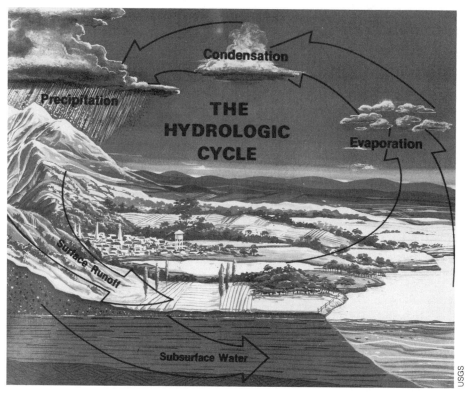

Fig. 2-1. The hydrologic cycle involves the flow of water from the ocean onto the land and back into the sea again.

The quickest route water takes back to the sea is by runoff in rivers and streams. Not only is this the most apparent part of the water cycle, but it is also the most important. Surface runoff brings minerals and nutrients to the ocean and cleanses the land. Acidic rainwater reacts chemically with metallic minerals on the surface, creating metallic salts that are carried in solution by rivers to the sea. This is also one of the reasons why the ocean is so salty. Rainwater also percolates into the ground, dissolves minerals from porous rocks, and transports them by groundwater.

The rainstorms that bring water to the Earth's surface are of vital importance. The continents receive about 25,000 cubic miles of water yearly. Some 15,000 cubic miles of it evaporates from lakes, rivers, aquifers, soils, and plants. Trees and other vegetation lose a great deal of water by *transpiration*, the evaporation of water from leaves and other plant parts. About 15 percent of all the moisture in the atmosphere comes from the land.

The remaining 10,000 cubic miles of water is stable runoff by rivers and streams. Much of this surplus water is lost by floods. The United States has about 3.5 million miles of rivers and streams; about six percent of the land adjoining these water courses is prone to flooding. Floods are natural, reoccurring events that are important for the distribution of soils over the land. During a flood, a river might change its course several times as it meanders to the sea, thus laying out a new flood plain. Flood plains provide level ground and fertile soil, so they also attract commerce. Rapid development of these areas without considering the flood potential often ends in disaster (FIG. 2-2).

PHOTO BY J.M. CHILDERS, COURTESY OF USGS

Fig. 2-2. Severe flooding at Fairbanks, Alaska on August 15, 1967.

In the last 50 years, the Federal Government has spent over $10 billion on flood protection projects. Most of these are man-made reservoirs, which even out the flow rates of rivers with a storage capacity that can absorb increased flow during a flood. The dams also provide hydroelectric power (FIG. 2-3), the cleanest form of energy. However, reservoirs also submerge valuable land areas, spurring furious water wars—especially in the dry western states. Without proper soil conservation measures, the accumulation of silt from soil erosion can also severely limit the life expectancy of a reservoir.

Most of the eastern third of the nation receives sufficient rainfall to support agriculture without the need for irrigation. In contrast, much of the western portion of the nation has a rain deficit that must be filled by irrigation (FIG. 2-4). Over 10 percent of the world's cultivated land is irrigated, requiring almost 1000 cubic miles of water. Irrigation permits crops to be independent of nature for their water. Furthermore, additional land can be cultivated and two or more crops can be grown in a single year.

Fig. 2-3. Hoover Dam and Lake Mead on the border between Nevada and Arizona.

Irrigation also has some serious drawbacks. Overuse of groundwater depletes aquifers, which can cause subsidence and severely limit the recovery of groundwater systems. Moreover, river water used for irrigation often has a high salt content. Poorly drained fields allow salt to build up in the soil, and every year thousands of once-productive acres are ruined.

THE CARBON CYCLE

The carbon cycle keeps our planet alive in the biological sense as well as the geological sense. The recycling of carbon through the biosphere makes the Earth unique. This

Fig. 2-4. An in-line type emitter that is used on a drip irrigation system in Imperial County, California. This system conserves water and energy and reduces soil erosion.

is evidenced by the fact that the atmosphere contains large amounts of oxygen. Without the carbon cycle, this oxygen would have long since been buried in the geologic column. Fortunately, plants replenish oxygen by utilizing carbon dioxide, a primary source of carbon for photosynthesis. Therefore, this system provides the basis for all life.

Presently, carbon dioxide makes up about 350 parts per million of the air molecules in the atmosphere, amounting to about 700 billion tons of carbon. It is one of the most important greenhouse gases; it traps solar heat that would otherwise escape into space (FIG. 2-5). Therefore, carbon dioxide plays a vital role in regulating the earth's temperature and major changes in the carbon cycle would profoundly affect the climate. In this respect, carbon dioxide operates like a thermostat, regulating the temperature of the planet. If too much carbon dioxide is removed from the atmosphere by the carbon cycle, the Earth would cool. If too much carbon dioxide is generated by the carbon cycle, the Earth would heat. Therefore, even slight changes in the carbon cycle could affect the climate and ultimately life.

The oceans play a major role in regulating the level of atmospheric carbon dioxide. In the upper layers of the ocean, the concentration of gases is in equilibrium with the atmosphere. The mixed layer of the ocean, the upper 250 feet, contains as much carbon diox-

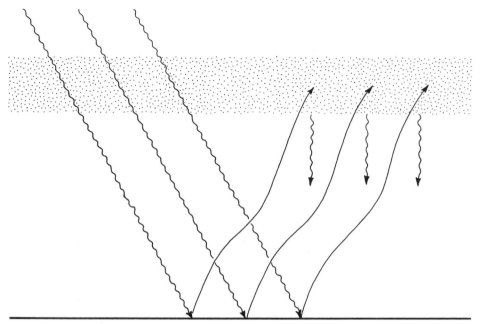

Fig. 2-5. The principle of the greenhouse effect. Incoming solar energy is converted to infrared radiation, which is absorbed by greenhouse gases (mainly carbon dioxide) and reradiated back to the ground.

ide as the entire atmosphere. The gas dissolves into the waters of the ocean mainly by the agitation of surface waves (FIG. 2-6). If no life was in the ocean, much of its reservoir of dissolved carbon dioxide would escape into the atmosphere, more than tripling its present content.

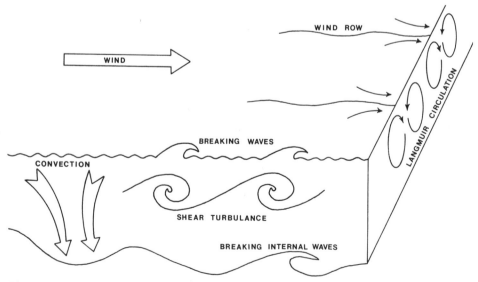

Fig. 2-6. Turbulence in the upper layers of the ocean induces temperature and nutrient mixing.

Fortunately, the ocean is teeming with life, and marine organisms use carbon dioxide (in the form of dissolved bicarbonates) to build their carbonate skeletons and other supporting structures. When the organisms die, their skeletons sink to the bottom, where they dissolve in the deep waters of the abyss, which holds by far the largest reservoir of carbon dioxide. The ocean contains about 60 times more carbon than the atmosphere, mostly in the form of dissolved bicarbonate.

Most of the Earth's carbon is stored in sediments on the ocean floor and on the continents. In shallow water, the carbonate skeletons build up thick deposits of carbonate rock, such as limestone (FIG. 2-7), which buries carbon dioxide in the geologic column. The burial of carbonate in this manner is responsible for about 80 percent of the carbon deposited

Fig. 2-7. Basal limestone of the Hawthorn Formation rests on Ocala limestone in Marion County, Florida. The man is standing on the contact between the two rock formations.

The Carbon Cycle **21**

on the ocean floor. The rest of the carbonate comes from the burial of dead organic matter washed off the continents. Half of the carbonate is transformed back into carbon dioxide, which returns to the atmosphere. Without this process, all the carbon dioxide would have been removed from the atmosphere in a mere 10,000 years. The loss of this important greenhouse gas would end photosynthesis and force all life into extinction.

The deep water, which represents about 90 percent of the ocean's volume, circulates very slowly and has a residence time of about 1000 years. It communicates directly with the atmosphere only in the polar regions, so its absorption of carbon dioxide is very limited in these areas. The abyss receives most of its carbon dioxide in the form of shells of dead organisms and fecal matter that sink to the bottom. The carbon dioxide is returned to the atmosphere by upwelling currents in the tropics. For this reason, the concentration of atmospheric carbon dioxide is greatest near the equator.

Volcanic eruptions are the final stage of the carbon cycle. Volcanic activity plays a vital role in restoring the carbon dioxide content of the atmosphere. The carbon dioxide escapes from carbonaceous sediments when they melt after being forced into the mantle at subduction zones near the edges of continents and volcanic island arcs (FIG. 2-8). The molten magma, along with its content of carbon dioxide, rises to the surface to feed volcanoes and midocean rifts. When the volcanoes erupt, carbon dioxide is released from the magma and is returned to the atmosphere. The cycle is complete.

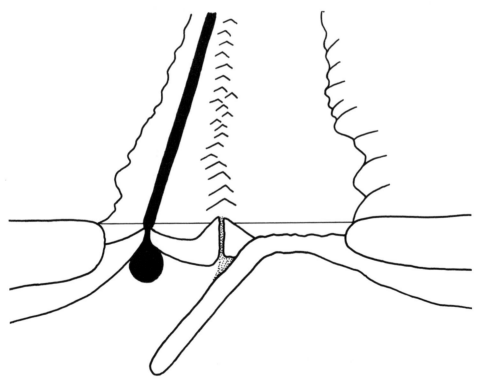

Fig. 2-8. An island arc, created by a subduction zone where two plates converge, is caught between colliding continents.

THE NITROGEN CYCLE

Nitrogen composes nearly 80 percent of the atmosphere and it is one of the major constituents of life. Carbon, nitrogen, and hydrogen are the essential elements for making proteins and other biological molecules. However, nitrogen is nearly an inert gas and it requires special chemical reactions for it to be utilized by nature. In order for it to combine with other substances a great deal of energy is required.

The nitrogen cycle is a continuous exchange of elements between the atmosphere and living organisms by the action of other organisms such as nitrogen-fixing bacteria. All methods of nitrogen fixation require a source of energy, which mainly comes from the Sun. The decay of organisms after death releases nitrogen back into the atmosphere, thus the cycle is completed.

Atmospheric nitrogen originated from early volcanic eruptions and the breakdown of ammonia (one nitrogen atom and three hydrogen atoms), which was a large constituent of the primordial atmosphere. Unlike most other gases, which have been permanently stored in the crust or have been replaced, the Earth still retains much of its original nitrogen. The nitrogen is easily transformed into nitrate, and dissolved in the ocean. There, denitrifying bacteria return the nitrate-nitrogen to its gaseous state. Otherwise, all atmospheric nitrogen would have disappeared long ago and the Earth would have only a fraction of its present atmospheric pressure.

Nitrogen oxides found in acid rain can be especially harmful to aquatic organisms. Nitrogen is a nutrient that promotes the growth of algae (FIG. 2-9). The algae blocks out sunlight and depletes water of its dissolved oxygen, which in turn suffocates other aquatic plants and animals. Nitrate levels, along with higher concentrations of toxic metals (including arsenic, cadmium, and selenium), have been greatly increasing. The main factors contributing to this increase are fertilizer and pesticide runoff, along with acid rain, which dissolves heavy metals in the soil. Acid rain also depletes the soil of some of its nutrients, including calcium, magnesium, and potassium. Some soils are so acidic that they can no longer be cultivated.

Oxides of nitrogen, produced in factory furnaces and by motor vehicles, absorb solar radiation and initiate a chain of complex chemical reactions. In the presence of organic compounds, these reactions form a number of undesirable secondary products that are very unstable, irritating, and highly toxic. High-temperature combustion yields nitrogen oxide and gaseous nitric acid. Because of their minute size, these particles, called *aerosols*, scatter light that normally would heat the ground. Instead, they heat the atmosphere and cause a temperature imbalance between the atmosphere and the surface, creating abnormal weather patterns.

Nitrous oxides are produced by the combustion of fossil fuels—especially under high temperatures and pressures (for example, those in coal-fired plants and internal combustion engines). The tall chimneys of coal-fired plants send huge amounts of nitrous oxides into the atmosphere. The gas eventually gets into the upper stratosphere, where it can break down the ozone that protects the surface from ultraviolet radiation.

Deforestation might also be a threat to the ozone layer by releasing nitrous oxide into the atmosphere. Clear-cutting timber allows soil bacteria to produce nitrous oxide, which is expelled into the atmosphere. The tremendous heat produced by the burning timber combines nitrogen and oxygen into nitrous oxide, and significant amounts of this

Fig. 2-9. Rafts of blue-green algae lying in a roadside pool in Indian Canyon, Duchesne County, Utah.

gas escape into the upper atmosphere. A continued depletion of the ozone layer with accompanying high ultraviolet exposures could reduce crop productivity and aquatic life—especially the primary producers at the very bottom of the food chain.

THE BALANCE OF NATURE

The Earth not only provides life with all the essentials it requires to survive, but life also appears to have made some changes of its own in order to live at its optimum. The *Gaia hypothesis,* named for the Greek goddess of the Earth, suggests that life is capable of controlling, to some extent, its own environment to maintain optimum living conditions. Life uses the atmosphere both as a source of raw materials, such as oxygen and nitrogen, and as a repository for waste products, such as carbon dioxide. In this manner, life has a direct input to the greenhouse effect. Thus, it appears that living organisms can regulate the climate.

From the very beginning, life seems to have followed a well-organized pattern of growth as it developed from simple to complex organisms. Originally, the atmosphere contained about as much carbon dioxide as it now has oxygen. Because the Sun was so much weaker than it presently is, the carbon dioxide was needed to maintain the Earth's temperature, otherwise it would have completely frozen over.

When green plants evolved, they gradually replaced carbon dioxide with oxygen. This is fortunate because at the same time the Sun was getting progressively hotter and large amounts of carbon dioxide were no longer needed in the atmosphere. If the carbon dioxide was not removed from the atmosphere, the Earth would have suffered the same fate as Venus, whose surface temperatures are hot enough to melt lead. Moreover, if the Earth had today's atmosphere in the beginning, it would have been as cold as Mars and the oceans would have frozen solid.

When oxygen reached higher levels, complex organisms began to evolve. When the level was near where it is today, the ozone screen made it possible for plants and animals to conquer the land. Very few places on Earth are truly devoid of life. It exists in the hottest deserts and in the coldest polar regions (FIG. 2-10). Life resides in lowest canyons

PHOTO BY W.B. HAMILTON, COURTESY OF USGS

Fig. 2-10. Daniell Peninsula, Antarctica.

and tallest mountains. It also lives in the deepest oceans and in the highest regions of the troposphere. Nor is life excluded from scalding hot springs or the deep underground.

Although the species most frequently encountered on the Earth's surface would seem to be the most influential force in shaping the planet, it is actually the unseen microbes, which constitute about 90 percent of the biomass, that have the greatest effect. They are morphologically simple creatures, but are biochemically diverse, highly adaptive, and essential for maintaining living conditions on Earth.

Eighty percent of the Earth's breathable oxygen is generated by photosynthetic single-cell organisms that thrive in the ocean. Microorganisms, such as bacteria, play a critical role in breaking down the remains of plants and animals for recycling nutrients. Surface plants depend on bacteria in their root systems for nitrogen fixation. Bacteria live symbiotically in the digestive tracts of animals and aid in the digestion of food. Simple organisms are the bottom of the food chain, upon which all life ultimately depends for its survival.

Living things also have the ability to store energy by combining carbon from carbon dioxide in the atmosphere with hydrogen from water to form *carbohydrates* (FIG. 2-11).

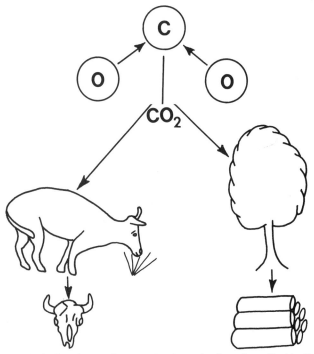

Fig. 2-11. The biological carbon cycle converts atmospheric carbon dioxide directly into vegetative matter or indirectly into animal matter, which reverts back to carbon dioxide upon decay or combustion.

Thick coal deposits, which can be thought of as buried solar energy because they originated as vegetative matter, and vast subterranean reservoirs of oil, which are cooked hydrocarbon molecules from once-living microorganisms, have been accumulating over eons. When fossil fuels are burned in automobiles and factories, the equation is reversed and carbon is recombined with oxygen, releasing carbon dioxide back into the atmosphere. In this manner, humans are short-circuiting the carbon cycle.

The combustion of tremendous amounts of fossil fuels, the pollution of the environment with toxic wastes, the destruction of forests and wildlife habitat, and the uncontrollable human population explosion places man in the unique position of greatly changing Earth in a comparatively short period of time. Thus, humans are the single most destructive force on the face of the planet.

3

Danger in the Air

AIR pollution has become a growing threat to the world's health and welfare because of the ever-increasing emissions of contaminants into the atmosphere (TABLE 3-1).

TABLE 3-1. Approximate Nationwide Emissions (in Millions of Tons).

SOURCE	CARBON MONOXIDE	PARTICULATES	SULFUR OXIDES	HYDROCARBONS	NITROGEN OXIDES
Transportation	92	1	1	12	10
Industry	9	12	29	13	15
Waste Disposal	3	1		1	
Other	5	1		4	
Total	109	15	30	30	25

Each day, an average adult inhales about 30 pounds of air, drinks 5 pounds of water, and eats 3 pounds of food. In an entire lifetime, this amounts to about 400 tons of air circulating through the lungs. With this figure in mind, it is not too difficult to understand the need for breathing clean air. In the United States, 60 percent of the population (about

140 million people) live in areas where the air does not meet the standards set by the 1970 Clean Air Act.

LIVING IN A HAZE

All foreign substances injected into the atmosphere by man-made and natural sources are air pollution. Humans are by far the greatest polluters. Since the industrial era began two centuries ago, our species has rivaled nature for the amount of toxic wastes and particulates disposed of in the atmosphere.

Natural pollutants include salt particles from ocean spray, pollen and spores released by plants, smoke from forest fires set by lightning strikes, wind-blown and meteoritic dust, and volcanic ash (FIG. 3-1). Volcanoes are perhaps the largest natural polluters in

Fig. 3-1. A large eruption cloud from Mount St. Helens, Skamania County, Washington on July 22, 1980.

the world. However, much of their ejecta is beneficial to the environment. In the Great Smoky Mountains of eastern Tennessee, pine sap reacts with sunlight and damp air to generate a natural hydrocarbon-type photochemical haze that is similar to the smog in big cities.

The load of particulate matter consisting of soot and dust suspended in the atmosphere at any one time as the result of human activity is estimated to be about 15 million tons and is rapidly escalating. Slash-and-burn agriculture, which is destroying millions of acres of forestland each year, is responsible for tremendous amounts of smoke in the troposphere.

Dust blown off newly plowed or abandoned fields has been on the rise, clogging the atmosphere and severely eroding the land (FIG. 3-2). Factory smokestacks emit huge

Fig. 3-2. A road adjacent to an unprotected cotton field is buried during a dust storm near Floydada, Texas on January 25, 1965.

quantities of soot and aerosols. Motor vehicle exhaust accounts for 50 percent of the particulates and aerosols injected into the atmosphere. In addition, the wheels of moving vehicles inject roadway grit into the air, prompting many municipalities to step up street-sweeping operations to reduce the persistent brown haze that hangs over their cities.

Around the country, thousands of tons of dangerous chemicals are released into the atmosphere by chemical companies. One of these substances, used in making pesticides, killed 3000 people in Bhopal, India during an industrial accident in 1985. Over 200 hazardous chemicals are vented into the air, and some industrial plants spew millions of pounds of known *carcinogens* (cancer-causing agents) into the atmosphere each year.

Although we have no conclusive evidence concerning the health hazards of these toxic pollutants on the population, many scientists are concerned about the long-term exposure to air contaminants. Several of these substances are rained out of the atmosphere and end up in soils, rivers, lakes, and the sea, where they are concentrated by chemical and biological factors.

The result of all these pollutants clogging the skies is a decrease in sunlight reaching the ground and a subsequent cooling of the surface. Sunlight striking airborne particulates also heats the atmosphere, causing a thermal imbalance and unstable weather. Possibly one reason why an increased level of atmospheric carbon dioxide has not yet shown a large upward trend in global temperatures brought on by the greenhouse effect is that it is offset somewhat by the cooling effects of man-made particulate matter in the atmosphere.

The skies over the United States also appear to be getting cloudier—especially during the past several years. A comparison of the number of cloudless days, on which an average of 10 percent or less of the daytime sky was obscured by clouds, fog, haze, or smoke, was made for 45 major cities during 1900-1936 and 1950-1982. The results indicated that the second half of this century had more cloudy days than the first half. Some cities, such as Los Angeles, California, had an increase in cloudiness from 20 to 30 percent or more. Apparently, air pollution is supplying the microscopic particles that aid in the condensation of cloud droplets.

THE DIRTY SKIES

Air pollution is classified into *primary pollutants*, which are emitted directly from primary sources, such as smokestacks and motor vehicle exhaust pipes, and *secondary pollutants*, which are emitted from chemical reactions occurring among the primary pollutants. Many reactions responsible for secondary pollutants are triggered by sunlight and are called *photochemical reactions*. Nitrogen oxides, produced by factory furnaces and motor vehicles, absorb solar radiation and initiate a chain of complex chemical reactions. In the presence of organic compounds, these reactions form a number of undesirable secondary products that are very unstable, irritating, and toxic.

For the past few decades, the release of cancer-causing chemicals into the atmosphere has been on the rise. Each year, thousands of tons of dangerous chemicals are released by factories around the world. Many of these substances are rained out of the atmosphere and end up in rivers, lakes, and soils, where they are concentrated to toxic levels.

Most of the air pollution that reduces visibility and harms plants and animals as well as man-made structures is composed of dry deposits. These airborne particles consist of unburned carbon, dust particles, and minute sulfate particles. The finest particles are mainly produced by chemical processes, resulting from the combustion of fossil fuels. This pollution is essentially acidic.

High-temperature combustion in coal-fired generating plants and internal combustion engines yields nitrogen oxide and gaseous nitric acid. These minute particles, *aerosols*, scatter light that would otherwise heat the ground. Instead, they absorb sunlight and heat the atmosphere. These aerosols cause a temperature imbalance between the atmosphere and the ground, causing abnormal weather.

Coarse atmospheric particles are derived mainly from the mechanical breakup of naturally occurring substances, such as the ash from volcanic eruptions and sediment suspended in the air by dust storms. Large particles of carbon, called *soot*, are produced by forest and brush fires (FIG. 3-3) and inefficient combustion of wood-burning stoves and fireplaces.

Fig. 3-3. A brush fire that raged out of control in Ventura County, California on August 23, 1972.

Under an atmospheric inversion, whereby a warm layer of air acts like a lid on top of cold air near the ground, the smoke can create a persistent haze during the winter. Even once-pristine areas, such as the Arctic tundra, have significant levels of sulfate-containing particles, which originated from distant industrial sources.

DILUTION IS THE SOLUTION

Air pollution is particularly hazardous to the immediate area if it is not diluted by the atmosphere. If the air is unstable with turbulent winds, smoke and exhaust fumes are carried upward by air currents, mixed with cleaner air aloft, and dispersed by upper-level winds. The wind speed has a direct effect, and the concentration of pollutants is cut in half if the wind speed is doubled (FIG. 3-4). Therefore, if the air into which the pollution is

Fig. 3-4. The dispersion of air pollution is only half as much with a wind of 5 miles per hour (top drawing) as with one of 10 miles per hour (bottom drawing).

released is not dispersed by the wind, air pollution increases. Thus, air pollution problems are associated with periods when winds are weak or calm, but seldom occur when winds are strong.

Once the pollutants are well mixed with the air, they react with other atmospheric substances, such as oxygen and water vapor, to produce a secondary type of pollution. This mixture creates photochemical smog, corrosive acids, and deadly poisons, such as ozone, mainly from motor vehicle exhaust. This smog scars the lungs just like cigarette smoke.

High-pollution days do not necessarily indicate an increase in the output of pollution, but rather the air into which the pollution is released is not dispersed by the wind, making the air more toxic. Stagnant air under a zone of high pressure allows little vertical mixing of the pollutants with the cleaner air above and air quality drops. During a *temperature inversion*, in which warm air overlies cooler air and acts like a lid to prevent upward movement, pollutants become trapped below (FIG. 3-5). As the global temperature rises,

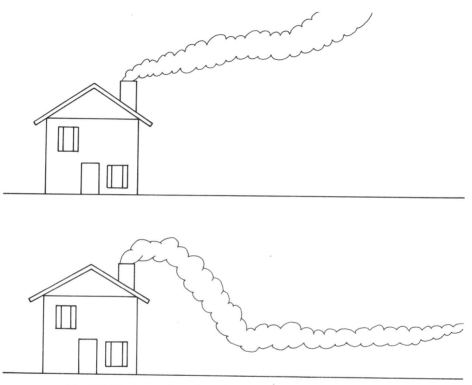

Fig. 3-5. During an inversion, air pollution remains near the ground.

the weather could shift to fewer and weaker storms, combined with a sluggish movement of air masses. So, the dirty air masses would remain hovering over industrial centers longer than they now do.

Geography also plays an important role in trapping air pollution. Although valleys are generally preferred sites for industrialization, they are also more likely to experience inversions that trap air pollution. In areas like Los Angeles, polluted air is caught in the valley and it cannot escape during a temperature inversion. The worst air pollution disaster occurred during a strong inversion that resulted in the great 1952 London smog, which killed 4000 people in an 11-day period.

In an effort to eliminate local air pollution from coal-fired plants, companies built their smokestacks high above the ground (FIG. 3-6) so that pollutants would readily mix with the turbulent air above and be carried aloft. Unfortunately, this only turned a local pollution problem into a regional one by allowing pollutants to travel for long distances,

Fig. 3-6. A large coal-fired generating plant near Muskogee, Oklahoma.

producing a host of problems downwind. Canada suffers from pollution from the industrialized areas of the Ohio River Valley. Sweden and Norway are constantly being bombarded by pollution from the heavily industrialized regions of Europe.

ACID PRECIPITATION

Human activity accounts for roughly 10 times more sulfur in the atmosphere than from natural sources. Sulfates, the leading constituents of acid rain, also cut visibility by upwards of 50 percent or more, leaving many parts of the nation in a persistent haze. The industrial era brought the combustion of high-sulfur coal and oil, along with the smelting of sulfide ores, particularly in the heavily industrialized regions of the Northern Hemisphere.

The combustion of sulfur, mostly in coal-fired furnaces, produces sulfur dioxide gas, which enters the atmosphere, combines with oxygen, and yields sulfur trioxide. This substance combines with moisture in the atmosphere to produce sulfuric acid. In addition, nitrogen oxides produced by high-temperature combustion gives rise to nitric acid. These acids mix with rainwater to produce an extremely corrosive acid rain.

The acidity levels of rain and snow indicate that in many parts of the world, especially in eastern North America and northwestern Europe (FIG. 3-7), precipitation has changed from a nearly neutral solution at the beginning of the industrial era to a dilute solution of sulfuric and nitric acid today. In the most extreme cases, the rain has the acid-

Fig. 3-7. Areas of heavy acid precipitation in the world.

ity of vinegar. Even in virtually unindustrialized areas, such as the tropics, acid rain occurs, mostly from burning rain forests.

Acid rain has been known to exist for many decades in the vicinity of large cities and industrial plants. This knowledge prompted the construction of tall smokestacks to disperse emissions high into the atmosphere and away from the cities. Unfortunately, the pollutants travel long distances, even crossing international borders. Coal-fired electricity generating plants in the Ohio River valley, for example, produce acid rain clouds that travel into eastern Canada, where the acid rain is destroying forests and aquatic habitats.

Lakes and streams in other parts of the world (FIG. 3-8), especially those that are not buffered by carbonate rocks, which help neutralize the acid, have become so acidic from acid rain runoff or polluted by toxic wastes that fish populations have been nearly decimated. Besides killing fish directly, acid rain also kicks the bottom out from under aquatic food chains and entirely changes the organic composition of the lakes. In the Adirondack Mountains of New York, 90 percent of the lakes that have acid levels below a pH of 5 are completely devoid of fish. The *pH value* is a logarithmic scale that ranges from 0 (the strongest acid) to 7 (a neutral solution) to 14 (the strongest base). In Canada, 14,000 lakes are threatened by acid rain. In Sweden, it is estimated that more than 15,000 lakes can no longer support fish populations because of acid rain.

Acid precipitation is especially harmful to aquatic organisms. The damage arises from nitrogen oxides in acid rain. Nitrogen acts like a nutrient that promotes the growth of algae, which blocks out sunlight and depletes the water of its dissolved oxygen. The loss of oxygen, which occurs when the algae die and are consumed by bacteria, suffocates other aquatic plants and animals.

Plants are also damaged by the adverse effects of acid on foliage and root systems, which is actively destroying the great forests of North America, Europe, China, and Brazil. The tree-ring width is decreasing and the mortality rate of red spruce in the eastern United States is increasing. Resorts and wilderness areas, such as those in the western United States, Norway, and West Germany are losing much of their natural beauty as a

Fig. 3-8. Areas in the United States where streams and lakes are highly acidified.

result of acid rain. Mountain forests are particularly at risk when covered by acid clouds because the bases of the clouds are usually much more acidic than the acid rain they produce.

Besides acid rain, is acid snow, acid fog, and acid dew. Although acid dew does not rival acid rain as an environmental hazard, it can be damaging and might play a significant role in harming trees. The acidic dewdrops are not usually harmful at night. However, evaporation after sunrise concentrates the acids, which could damage leaf surfaces. Acid dew forms when dewdrops absorb atmospheric nitric acid gas and sulfur dioxide, which is oxidized into sulfuric acid. It is also caused by dry deposition of acid particles and gases settling on wet surfaces.

Roughly 30 percent of the sulfur dioxide produced in the United States reaches the ground by way of dry deposition. It is suspected that dry deposits are just as destructive to the environment as acid precipitation. Sulfates are known to contribute most of the fine-particle mass over much of the eastern United States and many other regions. The sulfate particles are often highly acidic and could possibly damage materials and alter the acid-base balance as much as acid rain.

THE OZONE HOLE

Scientists using satellites to monitor the ozone concentration in the upper atmosphere have discovered that the ozone layer is rapidly being depleted. The ozone layer, which lies between 20 and 30 miles above the Earth, is only a trace constituent of the stratosphere, with a maximum concentration of only a few parts per million of the air molecules. If the ozone layer was concentrated into a thin shell of pure ozone gas surrounding the Earth at the pressure of one atmosphere, it would measure only about 1/8 inch thick.

Every September and October since the late 1970s, a giant hole (about the size of the continental United States) opens in the ozone layer over Antarctica, where 50 percent of the ozone disappears. During the record-setting year of 1987, the Antarctic ozone hole grew twice as large as the continent itself. A similar ozone hole has been discovered over the Arctic as well. The yearly formation of the Antarctic ozone hole could alter the region's marine ecology by allowing high levels of ultraviolet radiation to destroy *phytoplankton*, which form the base of the ocean food web that other sea life depend on for survival.

Furthermore, these regions export their depleted ozone air and ozone-destroying chemicals to the mid-latitudes when the ozone holes break up. As the holes grow larger, which they seem to be doing every year (FIG. 3-9), chances are that more ozone-depleted air will leak into the lower latitudes and cause damaging effects from ultraviolet radiation. For example, in southern Australia, ultraviolet levels climbed by about 20 percent for a few days following the breakup of the Antarctic ozone hole in December 1987.

The ozone shield over the United States is eroding twice as fast as was previously thought. Long-term records show that ozone levels in the high northern latitudes have dropped about 5 percent over the last two decades and they might drop another 5 percent by the end of this decade. The ozone depletion is strongly believed to be caused by man-made chemicals. Unfortunately, even if the chemical emissions completely ceased, the ozone layer would continue to diminish for at least another century because that is the time required to cleanse the upper atmosphere of ozone-destroying chemicals.

Chlorine monoxide, which exists in atmospheric pollution and destroys ozone molecules, was found to be 100 times the normal level in the Antarctic stratosphere. Polar stratospheric clouds composed of frozen water and nitric acid crystals help chlorine destroy the ozone layer by chemically reacting with the free oxygen atoms that compose ozone molecules. The Arctic ozone hole is less well-defined, compared to the one in the Antarctic, because temperatures are not cold enough to allow the cloud crystals to form.

Ozone is produced in the upper stratosphere by the absorption of solar ultraviolet radiation by oxygen molecules. When the chemical bond is ruptured, free oxygen atoms are liberated. These atoms attach to oxygen molecules to produce unstable ozone molecules composed of three oxygen atoms. The ozone then decays back to an oxygen molecule and an oxygen atom within the ozone layer (FIG. 3-10).

Certain chemicals released into the stratosphere directly compete for the free oxygen atoms in the ozone layer and thereby interfere with ozone production. Furthermore, the mechanisms by which ozone is destroyed are based on chemical chain reactions, in which one pollutant molecule can destroy many thousands of ozone molecules before it falls to the lower atmosphere, where it can no longer do any harm.

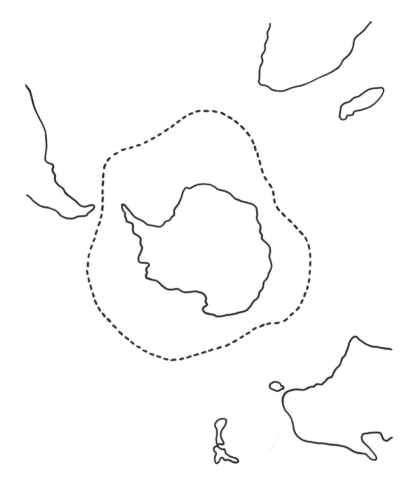

Fig. 3-9. If the ozone hole continues to expand annually, it might eventually drift into continents that surround Antarctica.

Chlorofluorocarbons (CFCs) and nitrous oxides are the primary culprits destroying ozone. CFCs are used in refrigerators and air conditioners and they escape into the atmosphere when these units are manufactured or damaged. CFCs are also used as propellants in spray cans and in the manufacture of foam plastics. CFCs are also used as industrial solvents; evaporation and spillage releases these chemicals into the atmosphere. Some 50 nations have signed a protocol that bans CFCs by the end of the century. Unfortunately, by that time, the ozone layer might already have been thinned to dangerous levels.

Ozone-destroying nitrous oxides are also produced by the combustion of fossil fuels—especially under high temperatures and pressures, such as those found in coal-fired plants and internal combustion engines. Substantial amounts are also produced by forest fires. Even the Space Shuttle releases ozone-destroying chemicals from its booster rockets that are equal to the destruction of 60,000 refrigerators every time it is launched.

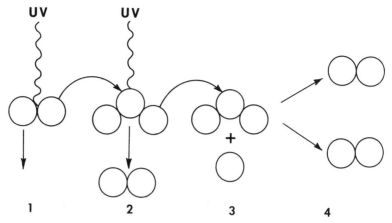

Fig. 3-10. The life cycle of an ozone molecule. 1) Ultraviolet (UV) radiation splits an oxygen molecule into 2 oxygen atoms. 2) One of these atoms combines with another oxygen molecule to create ozone. The ozone traps UV, liberating an oxygen molecule and an oxygen atom, to reform as ozone. 3) An addition of another oxygen atom then creates 4) two new oxygen molecules.

The variation of ozone concentration in the ozone layer is affected by the seasons, latitude, and strong weather systems, which can penetrate into the stratosphere. If ozone descends to the lower levels of the atmosphere, it is destroyed before it reaches the ground. This destruction is fortunate because ozone is also highly toxic near the surface, irritating eyes and lungs.

Ozone plays a vital role in shielding the surface from the Sun's harmful ultraviolet radiation. Without this shield, life could not exist on land or in the surface waters of the ocean. A slight increase in ultraviolet rays can increase health hazards, such as skin cancer, cataracts, and weakened immune systems, and increase harm to plants and animals. It can also exacerbate serious pollution problems, such as smog and acid rain. A continued depletion of the ozone layer, with accompanying high ultraviolet exposures, could destroy the primary producers, upon which all life on Earth ultimately depends for its survival.

4

Muddying the Waters

ALL of the fresh water in the world's rivers and lakes represents only about 0.01 percent of the total amount of water on the planet. Three-quarters of all the Earth's fresh water is locked up in glacial ice, which contains enough water to fill the Mediterranean Basin 10 times over. The rest exists in fresh water lakes, streams, aquifers, and living organisms. Only a small fraction of the fresh water supply is available for human needs, however, and most of it is used for agriculture. Unfortunately, much of the water is highly polluted, resulting in the deaths of millions of people from water-related diseases.

DUMPING IN THE SEA

Despite environmental pressures, industrial toxic chemicals are still making their way into rivers and streams—either deliberately or accidentally in industrialized nations (FIG. 4-1). Hospital wastes, including used syringes and contaminated dressings, which could cause serious diseases, are dumped into the sea and wash up on shore. Untreated or only partially treated sewage is dumped directly into the ocean because sewage-treatment plants are seriously overloaded.

So much waste is generated by large cities that the only place left to put it is into the sea. The cost of land disposal of toxic waste is escalating so greatly that coastal metropolitan areas are forced to dump industrial waste and raw sewage directly into the ocean. As a result, beaches in many parts of the world are unsightly and unsafe for swimming.

Rivers and coastal waters have become dumping grounds for millions of tons of toxic wastes every year. Some of these toxic pollutants are powerful carcinogens and muta-

41

Fig. 4-1. Water pollution in the Cumberland River, Nashville, Tennessee on May 12, 1970.

gens. Many more are nonbiodegradable and persist in the environment for extremely long periods. Ocean currents often return the wastes to shore; other wastes are concentrated between thermal layers and ocean fronts, where some of the most productive fishing grounds lie (TABLE 4-1).

TABLE 4-1. Productivity of the Oceans.

LOCATION	PRIMARY PRODUCTION TONS PER YEAR OF ORGANIC CARBON	PERCENT	TOTAL AVAILABLE FISH TONS PER YEAR OF FRESH FISH	PERCENT
Oceanic	16.3 billion	81.5	.16 million	0.07
Coastal Seas	3.6 billion	18.0	120.00 million	49.97
Upwelling Areas	0.1 billion	0.5	120.00 million	49.97
Total	20.0 billion		240.16 million	

The United States generates about 160 million tons of solid waste per day. Much of that ends up on the bottom of the ocean, where it becomes a menace to life. More than 250 dolphins washed ashore along the Atlantic Coast during the summer of 1987. Coastal sewage and industrial toxins were a prime suspect in their deaths.

Hydrocarbon chains, called *surfactants*, coat the surface of the ocean with a thin film that interferes with the transfer rates of gas and water vapor between the ocean and the atmosphere. One example is ordinary soap, which is a "dry" surfactant because it mostly rides on the surface of the water. This type of surfactant is rare, except in areas of man-made pollution—especially oil spills, which produce a thin, suffocating film that rides on the surface.

The largest rivers of North and South America empty into the Atlantic Ocean. The Atlantic is smaller and shallower than the Pacific, so it is also saltier. Along the east coast of the Americas, the sea lies on a wide and gently sloping continental shelf that extends eastward over 60 miles, reaching a maximum depth of 600 feet at the shelf edge. In contrast, along the west coast, the water descends rapidly to great depths just a short distance offshore. The ocean off the east coast is dominated by freshwater discharges from the great coastal plain estuaries from the St. Lawrence River in the north to the great Amazon River in the tropics. In marked contrast, the west coast is dominated by a periodic upwelling of bottom water from the Oregon coast to the coast of Peru.

The coastal seas of the world are among the most fragile and sensitive environments (FIG. 4-2). Laws passed to regulate wastes that can be discharged into the oceans were often based on inadequate knowledge of the sea. Therefore, compliance with those laws might do little to make the ocean cleaner. Some of the changes that human activity have

U.S. ARMY CORPS OF ENGINEERS

Fig. 4-2. Beach wave erosion of Grand Isle, Louisiana from Hurricanes Danny and Elana on August 30, 1985.

wrought in the ocean environment are irreversible, such as damming rivers, which lessens discharge into the sea, and building ports at the mouths of estuaries, which permanently changes the patterns of flow and alters the coastal habitat.

The world's oceans appear to be suffering from a dangerous decrease in vitality. The drop in marine species is much larger than if it was caused by chemical pollution alone. The decline is also caused by mechanical destructions, such as dynamite fishing, fishing in spawning grounds, using fine mesh fishing nets that trap the young and the adults, diverting rivers, filling marshes, and other destructive activities. If the attitude of exploiting the oceans for short-term gains does not change, the world faces a serious catastrophe.

No single ocean is healthier than any other because the currents make differences temporary. For this reason, DDT was found in the livers of penguins as far away as the Antarctic, where pollution was thought to be nonexistent. The pollutants in the vast cesspool of the Mediterranean Sea, where 20 percent of all beaches are unsafe for swimming, will eventually pollute the rest of the ocean.

The same is true for the Caribbean, the North Sea, the Gulf of Finland, and other heavily polluted seas. Although rivers and enclosed or semienclosed seas are presently more polluted than the open ocean, that situation could change in another decade or so. Even the middle of the Pacific Ocean, which was once thought to be pristine, is polluted with 100,000 particles per square mile.

HAZARDOUS OIL SLICKS

Oil spills are by far the most damaging of all coastal and river pollution. Increasing demand for offshore oil, collisions and groundings of oil tankers, attacks on oil tankers by belligerent nations, and deliberate dumping of oil into the sea as a form of environmental terrorism has led to disastrous ecological consequences. Every year, up to 25 million barrels of oil are spilled into the world's oceans. An estimated 1.5 million barrels of oil enter the ocean from natural seeps each year. However, this amount might easily be 10 times larger.

The number of oil spills are increasing steadily as consumption rises. A major oil-well blowout at Santa Barbara, California in January 1969, spilled 10 billion gallons of crude oil into the Pacific Ocean during the first 100 days, resulting in disastrous local ecological consequences. The oil spill prompted the United States Congress to pass legislation to limit industrial pollution of the nation's waters. The regulations required permits for discharging wastes into public waters and outlined limitations and monitoring requirements.

Nevertheless, major oil spills continued to occur. The December 1976 *Argo Merchant* spill off Nantucket, Massachusetts (FIG. 4-3) called for a reexamination of oil-spill contingency planning to protect productive fishing grounds. Oil from the 1978 wreck of the tanker *Amoco Cadiz* is still interfering with fish reproduction around the coast of Brittany off of France. The fouling of Texas beaches from another major oil-well blowout in the Gulf of Mexico in June 1979 continued to dramatize the dangers of offshore drilling. The 1986 spill on the Caribbean coast of Panama sent 50,000 barrels of crude oil into the sea and killed corals, mangroves, intertidal seagrasses, and other species that depended on these areas for their livelihood.

Fig. 4-3. The December 19, 1976 Argo Merchant *oil spill off Nantucket, Massachusetts.*

The spillage of massive quantities of oil in the Persian Gulf as a result of the eight-year Iran-Iraq war in the 1980s and of the 1991 Persian Gulf war continues to endanger the entire ecology of the Persian Gulf region and they could have serious consequences for many decades to come. Over one million barrels of oil was deliberately spilled into the Gulf by the Iraqi army soon after the beginning of the Persian Gulf conflict, ostensibly as a defensive move to prevent an Allied amphibious assault on Kuwait.

The grounding of the supertanker *Exxon Valdez* in Alaska's Prince William Sound in March 1989 was the nation's worst oil spill. The tanker released 250 thousand barrels of North Slope crude oil, generating an oil slick that covered a large portion of the Sound. Cleanup efforts were hampered by a lack of equipment and bad weather. Therefore, only about 10 percent of the lost oil was recovered out to sea. The rest washed up on nearby shores and soiled over 3000 miles of shoreline (FIG. 4-4). The spill will have a lasting effect on the local ecology and economy. The incident led to an examination of new regulations that will require oil tankers that operate in U.S. waters to be double-hulled, which could cut the amount of accidental spillage by 50 percent.

DANGER UNDERGROUND

More than 50 percent of the American people depend on groundwater for domestic use and irrigation (FIG. 4-5). Americans pump 100 billion gallons of water per day from aquifers. Most of the water comes from wells less than 300 feet deep. Many midwestern and western states draw more than half of their water from the ground. In several million rural households, aquifers are the only source of water. Therefore, contamination in

Fig. 4-4. Exxon workers use high-pressure cold water to rinse crude oil off a beach in Prince William Sound, Chugach National Forest, Alaska.

Fig. 4-5. Heavily irrigated areas in the United States.

these areas could be catastrophic because the surface water cannot satisfy domestic, industrial, and agricultural needs.

Toxic chemicals are leaching out of thousands of landfills throughout the United States. Agricultural pesticides and fertilizers are penetrating into the ground, and contaminants are percolating through layers of soil into groundwater aquifers (FIG. 4-6). Sub-

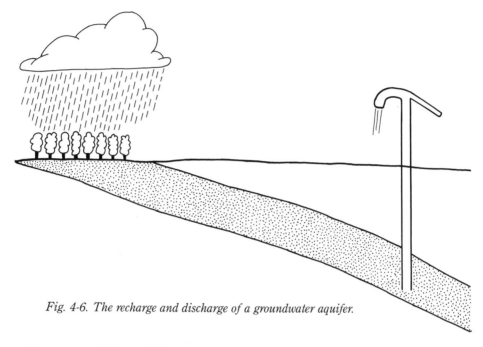

Fig. 4-6. The recharge and discharge of a groundwater aquifer.

surface water, which underlies millions of square miles of the country, an estimated 65 quadrillion gallons, is becoming increasingly polluted. The continental movement of subsurface water is extremely slow, taking as long as one million years.

Man-made organic chemicals, heavy metals, pesticides, and other toxic substances seep into the groundwater system from landfills, buried gasoline tanks, septic systems, radioactive waste sites, farms, mines, and numerous other sources. These pollutants might contaminate the nation's aquifers to such an extent that in the ensuing years, 25 percent of the nation's groundwater might be unusable. The groundwater problem has become the primary environmental challenge of the century. However, cleaning up the contamination will be expensive, difficult, and sometimes impossible.

Many thousands of wells across the nation have been contaminated by highly concentrated chemicals that spread through aquifers and exceed the Federal safe drinking water limits. Solvents and other chemicals from the manufacturing sector are leaking from buried storage tanks into the water supplies of several communities. Landfills that contain hazardous wastes are contaminating aquifers and forcing the shutdown of nearby water wells.

Waste lagoons and settling ponds are filled with toxic substances that end up in the groundwater. Although it has yet to be proven that small doses of toxic chemicals are harmful to humans, Federal regulators have assumed that long-term exposure to even minute quantities of most organic chemicals endangers the public's health. As much as 10 percent of the groundwater supply across the nation has already been contaminated. However, this percentage might be just the tip of the iceberg; contamination becomes greater when slow-moving patches of pollutants advance on populated areas. An increase in groundwater use will also worsen the problem because pumping speeds the flow of water in an aquifer and thereby increases the flow of the contaminants.

RUNNING OFF THE FARM

Perhaps as much as 30 percent of the global cropland is losing soil at a rate that is undermining any long-term agricultural productivity. The eroded soil is carried off the farm and into rivers that empty into the sea, where it ends up on the bottom of the ocean. Erosion rates are determined by the amount of precipitation, the topography of the land, the type of rock and soil, and the amount of vegetative cover.

In the United States, eroding cropland is costing the nation nearly one billion dollars annually from polluted and sedimented rivers and lakes (FIG. 4-7). Erosion severely limits the life expectancy of dams erected for water projects. The best way to control silt

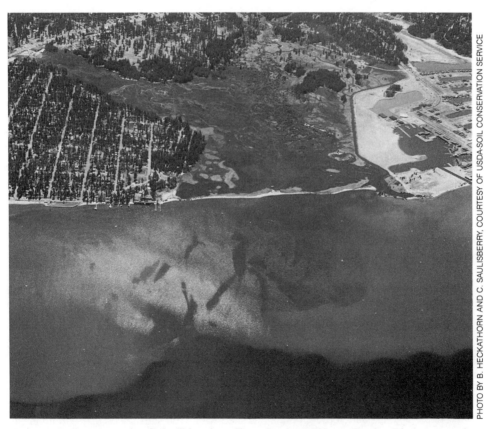

PHOTO BY B. HECKATHORN AND C. SAULISBERRY, COURTESY OF USDA-SOIL CONSERVATION SERVICE

Fig. 4-7. Sediment entering Lake Tahoe from Trout Creek and Upper Truckee River as a result of construction activities.

buildup is to adopt effective soil-conservation measures in the watershed so that less topsoil is lost.

Over 10 percent of the world's cultivated land is irrigated, requiring 600 cubic miles of water annually. Irrigation has many advantages as well as several drawbacks. Most river water that is used for irrigation has a high salt content. If fields are not drained properly, the buildup of salt in the soil can ruin the land and cause crops to become stunted or die. Tens of thousands of acres of once fertile land are destroyed by this process annually. Estimates indicate that by the end of the century, over half of all irrigated land will be rendered useless as a result of salt accumulation (FIG. 4-8).

Fig. 4-8. Areas affected by salt buildup in soils in the United States.

Irrigation, which not long ago turned vast stretches of the western desert of the United States into the world's most productive farmland, is now ruining hundreds of thousands of acres. It is also polluting marshes, rivers, lakes, and estuaries in California, Colorado, and other western states. Selenium that leached from soil on irrigated farms in the San Joaquin Valley of California has caused deformities in nearby waterfowl, including twisted beaks, stubs for wings, and no eyes. Contaminated birds lay thin-shelled eggs that can easily break, threatening a whole generation of waterfowl.

Irrigation water gradually degrades the land by the buildup of salts, including sodium, calcium, and magnesium chlorides. By the turn of the century, at least 30 percent of California's farmland could be destroyed by salt. Lower water availability in the spring and summer could also dramatically reduce crops. Meanwhile, the land in some regions contains selenium, arsenic, boron, and other naturally occurring poisons that pollute the runoff.

With most of the good irrigated land already in production, farmers can ill afford to cultivate the land until the salts build up and it has to be abandoned. The problem is much worse in arid regions because of the high natural salt content of the soil and low amounts of rainfall, which is needed to flush out the excess salts. Agricultural chemicals, such as fertilizers and pesticides, are carried off by the drainwater, which ends up in streams and rivers and finally in the ocean. If the concentrations are high enough, the chemicals can kill fish and other marine life.

ACCIDENTAL SPILLS

Offshore drilling and shipping accidents are the primary ways that 20 billion tons of dissolved and suspended matter reach the ocean, where their most dramatic effect is on the coastal zone. Many toxic substances that are diluted to supposedly safe levels in the ocean, lakes, and streams are concentrated by biological activity. At the bottom of the food chain, toxins accumulate in primary producers, which are eaten by fish and other aquatic life, some of which are a major dietary source for humans (FIG. 4-9).

Mercury poisoning of fish by industrial waste in rivers in Japan and DDT poisoning of penguins in the Antarctic worked through such a process. Some of the mercury is natural and some enters the water as fallout from air pollution created by incinerators, smelters, and coal-fired power plants. Fish are relatively unaffected by low levels of mercury. However, other animals, including humans, that eat the fish are at risk from mercury poisoning. Many toxic chemicals have spread throughout the world's marine ecosystem, in part through the gradual accumulation in the food chain.

One of the most serious accidental spills resulted from a fire at a Swiss chemical factory in Basel on November 1, 1986. About 30 tons of fungicides, pesticides, and other agricultural chemicals were washed into the Rhine River when firefighters attempted to put out the blaze, producing one of Europe's worst pollution incidents of the century. On one stretch of the upper Rhine from Basel, Switzerland, to Karlsruhe, Germany, over 100 miles to the north, the Rhine was nearly totally devoid of life, and at spots along the river, a purple sludge coated the river banks.

Huge populations of fish and eels died as the 25-mile-long chemical slick swept down the river, causing ecological damage that could take a decade or more to repair. At many spots along the 820-mile river, which is one of Europe's major waterways, supplies of drinking water had to be cut off and flood gates were closed to protect tributaries from contamination. Along the Rhine is one of Western Europe's most heavily industrialized regions. The river ultimately empties into the already highly polluted North Sea, where in 1988 more than 7000 seals died from industrial poisons. The accident substantially set back efforts to clean the Rhine, which had been ongoing for more than a decade.

The southern Volga River in the Soviet Union is on the brink of disaster because of the 300 million tons of solid waste and 5 trillion gallons of waste water that are dumped into it each year. Furthermore, 6 trillion gallons of water are drawn out of the river for use in agriculture and industry. Many other rivers, especially in Eastern Europe, where little or no environmental regulations are in place, have equally dismal conditions.

The Kattegat Sea between Sweden and Denmark is rapidly dying. The sea is so polluted and starved of oxygen that it is losing the capacity to support marine life. The first signs that the Kattegat was endangered came in the 1960s, when lobsters disappeared from the southern part of the sea. By the mid-1980s, thousands of dead fish were wash-

Fig. 4-9. The harvesting of catfish from a pond near Tunica, Mississippi.

ing up on the beaches of Germany, near the Danish border. Fishermen also reported large hauls of dead fish.

A Swedish research vessel found a severe lack of oxygen at all depths below 75 feet, and fish samples showed that the situation was extremely serious. The sea contains pollution from Denmark and southern Sweden, mainly pesticides and fertilizers draining off farmland. The Kattegat situation is difficult from an ecological standpoint because it is where the waters of the Baltic meet the much saltier currents of the North Sea, which itself is heavily polluted.

Far more alarming is that these conditions are not an isolated incidence; other regions might be suffering a similar fate as well. For example, the Aral Sea in the USSR can no longer be fished and the Sea of Asov might suffer the same fate as a result of

Fig. 4-10. A system of groins on Lake Michigan trap sands that move laterally along the beach to provide protection from waves.

pesticide runoff. The Baltic Sea is being heavily polluted by the Warsaw Pact countries, creating serious problems for Finland, Sweden, and Denmark.

Even the world's most pristine lakes, such as Baikal, the deepest and largest in the world by volume, are being threatened by pollution. The rising, polluted waters of the Great Lakes threaten beaches and coastal homes (FIG. 4-10). Toxic pollutants, including DDT and PCBs, rain directly into the lakes or run off contaminated areas on shore. These toxins are the leading environmental issue for the Great Lakes because 100 years or more are required for their waters to drain into the Atlantic Ocean.

5

The Dump Site
Dilemma

THE disposal of our mounting piles of garbage generated by our modern throwaway lifestyle remains as one of the most perverse problems we face as the century closes. Landfills in most major cities are overflowing and there is simply nowhere else to put the trash. Much of the garbage is not biodegradable, such as plastics, which remain in the environment for long periods. Even decades-old newspapers have survived in readable condition in landfills.

OUR THROWAWAY SOCIETY

Every day, Americans churn out 160 million tons of garbage—roughly 3.5 pounds of trash per individual. That number is expected to rise to nearly 200 million tons by the end of the century. If lined up bumper to bumper, a string of garbage trucks hauling the nation's annual waste could reach halfway to the moon.

Most of the waste is trucked to already overflowing landfills and buried. Often toxic substances leach out of the landfills and contaminate nearby water wells, requiring expensive treatment. As a result of the high cost of toxic waste land disposal, coastal metropolitan areas are dumping municipal and industrial wastes directly into the sea. Much of the waste that washes up on beaches comes from overburdened sewage-treatment plants, accidental spills from garbage barges, and lack of winds to disperse the flotsam. In addition, syringes and blood vials from hospital wastes that were illegally dumped into the sea and washed up on shore might contain infectious diseases, including AIDS.

Many toxic pollutants are powerful carcinogens and mutagens. Some are nonbiodegradable and persist in the environment for long periods. Ocean currents bring the wastes back to shore, making beaches unsightly and unsafe for swimming. Wastes are also concentrated between thermal layers and ocean fronts. These areas contain some of the most productive fishing grounds. Furthermore, the meandering currents of the Atlantic Gulf Stream (FIG. 5-1), which are often laden with fish, sweep directly over the dump sites.

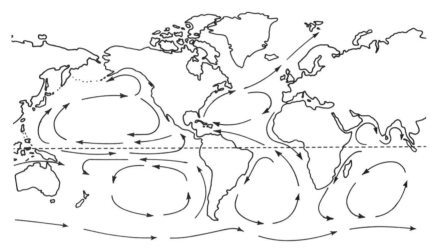

Fig. 5-1. The major ocean currents. The Gulf Stream is a warm current that originates in the Caribbean, passes the eastern coast of North America, and crosses the North Atlantic.

Recycling is one way out of this dilemma, and more than 75 percent of the municipal solid waste is recyclable material. However, implementing the process on a national scale is difficult. One of the problems is that some industries refuse to use secondary materials. Recycling has little economic incentive when alternatives to waste disposal, such as incineration abound. Unfortunately, incineration, turning trash into ash, has enormous pollution problems of its own. However, incineration can be minimized or avoided altogether through aggressive recycling.

Reducing the amount of garbage we generate requires changes in the American consumer ethic. These changes might include such things as higher taxes on packaging, banning certain unrecyclable plastics and throwaway products, and instituting standards for making products last longer. Furthermore, industries need assurances that the supply of recyclable raw materials is abundant and reliable. Tax incentives might also encourage industry to use recyclable materials.

States could pass laws that require deposits on beverage containers (FIG. 5-2), taxes on nonrecyclable packaging, and bans on certain materials. Individuals can help out by composting food and yard scraps, buying or selling second-hand merchandise, avoiding durable goods that do not last or waste energy, and using recycled materials whenever possible. These steps can be taken without requiring major changes in lifestyle. The alternative, however, is a continued reliance on incineration and its serious pollution problems.

Fig. 5-2. Aluminum beverage cans are emptied into a crushing and baling machine at the Pensacola Naval Air Station, Florida.

OVERFLOWING LANDFILLS

We are quickly running out of room to put our garbage. This enormous problem was brought to light during the summer of 1987, when a garbage barge from Long Island Sound attempted to offload its 10,000 tons of waste at six states and five different countries. After traveling 6,000 miles, it was forced to return to New York because no one wanted its cargo.

New York hosts the largest garbage dump in the world, containing enough refuse to fill the Panama Canal twice over. Once a pristine wetland on Staten Island, the landfill covers an area of 2,000 acres and receives about 26,000 tons of trash from New York City daily. As the trash heap continues to mount, it will some day have the dubious distinction of being one of the tallest points on the East Coast. It is also an environmental disaster, as two million gallons of *leachate* (liquid garbage) leaks into surrounding waterways every day. The concentration of methane gas is large enough to provide natural gas for 10,000 homes (FIG. 5-3).

Americans presently generate twice the garbage that they did in 1960 and twice as much per capita as other industrial countries, forcing the closing of landfills around the nation. Thousands more are rapidly nearing capacity and will be closed forever. Further-

Fig. 5-3. An anaerobic digestion plant at Pompano Beach, Florida uses urban solid waste and sewage sludge to produce a methane-rich gas that can be upgraded to commercial natural-gas quality.

more, few other places are available for the trash because of dwindling dump space and the not-in-my-backyard attitude of the local citizenry.

Most states in the highly populated Northeast and Midwest have less than a decade of dump space left. Old dumps, which receive 80 percent of all garbage, are rapidly filling up, and most will be forced to close in the next five years. Unfortunately, as the dump sites diminish, the garbage will continue to grow by upwards of 20 percent in this decade, if trends continue.

Paper is the largest single commodity in landfills, taking up as much as 50 percent by volume. Plastics, glass, metals, other inorganic materials, food wastes and yard wastes make up about 40 percent of the waste volume. Construction materials compose about 10 percent of the contents of landfills. The garbage does not decompose, but is well preserved because the contents are tightly packed and completely covered and exposed to little light or moisture. Some food debris and yard waste does slowly degrade, but the process can require decades.

As the nation's landfills are overflowing, waste disposal companies are looking for ways to dump unwanted garbage on Indian reservations. Indian tribes have been promised large advance payments to turn their land into vast dumps for household trash and toxic waste. Also included are proposals for building hazardous waste and nuclear waste disposal plants on Indian reservations. One reason why garbage companies are looking at Indian lands is that environmental regulations are less rigid than elsewhere. Many Indian tribes, however, refuse to poison their sacred lands with white man's refuse.

One escape from the garbage disposal dilemma is to build costly waste incinerators and possibly even use the heat to generate electricity. Unfortunately, one problem (waste disposal) is solved, while another problem (air pollution) is created. Thousands of tons of pollutants, including toxic dioxin, would be emitted into the air each year. Every 100 tons of trash generates 30 tons of ash, often laden with heavy metals, which qualifies it as hazardous waste, creating another garbage disposal problem. Furthermore, the incineration of wastes will exacerbate the greenhouse effect by exhausting huge amounts of carbon dioxide into the atmosphere.

A more acceptable solution to the growing garbage disposal problem is recycling, along with a concurrent reduction of convenience packaging that feeds our wasteful throwaway society. Recycling keeps trash out of landfills, it does not generate pollution, and it reduces the need to mine or harvest new raw materials, thereby saving the environment. Recycling will also reduce the need for incineration and the serious pollution problems that it entails.

TOXIC WASTE DISPOSAL

In May 1980, 400 homes adjacent to the Love Canal toxic waste dump in Niagara Falls, New York had to be permanently evacuated and the neighborhood fenced off as a result of a heavy contamination of toxic chemicals. The city itself, which is a chemical industrial center, was also polluted with substances leaking from other hazardous waste dumps. Comparable levels of pollution exist in many other urban areas, which might create adverse health effects.

Hazardous wastes, including man-made organic chemicals, heavy metals, pesticides, and other toxic substances, are seeping into the ground from landfills, buried gasoline tanks, septic systems, radioactive waste sites, farms, and mines (FIG. 5-4). The sources of pollution are so diverse that often it is difficult to determine the major cause of groundwater contamination. Routine monitoring of industrial waste lagoons and landfills reveal that the chemicals are not being contained and end up in nearby water wells, forcing many residents across the country to rely on bottled water.

The dumping of toxic wastes into the ocean is an insidious and potentially serious problem. Approximately eight million tons of toxic wastes are dumped into rivers and coastal waters each year. The long list of toxic substances includes chlorinated hydrocarbons, benzenes, trichloroethylene, toluene, polychlorinated biphenyls (PCB's), dioxin, solvents, organic chemicals, pesticides, fertilizers, fibrous asbestos, and heavy metals. Some of these toxic pollutants are powerful carcinogens and mutagens. Many are nonbiogradable and persist in the environment for extremely long periods.

Even minute amounts of these pollutants (below detectable limits) are dangerous. Toxic pollutants concentrate to lethal levels as they progress up the food chain (FIG. 5-5). At the base of the marine food chain are phytoplankton, which act as concentrators of pollutants. These are eaten by zooplankton, which in turn are fed upon by crustaceans and fish, some of which ends up on people's dinner tables.

Fig. 5-4. Mining operations often bring toxic substances to the surface that were buried deep in the ground. These substances can pollute groundwater aquifers and streams.

As a result of the escalating cost of land disposal of toxic wastes, many coastal metropolitan areas are forced to dump municipal and industrial wastes directly into the sea. Federal funds are not available to meet the over $100 billion needed for the nation's cities to dispose of their wastes. Therefore, the pressure on ocean dumping continues to rise.

Often, untreated sewage is dumped directly into the ocean, because sewage treatment facilities are overloaded and funds are unavailable for improvements. Besides human effluent, which itself is extremely toxic and full of pathogens, other municipal wastes, such as used crankcase oil, toxic wastes, and hospital wastes (including contaminated dressings and used syringes), end up in metropolitan sewage systems with the potential to damage the environment and spread diseases. As a result, beaches in many parts of the world must be closed.

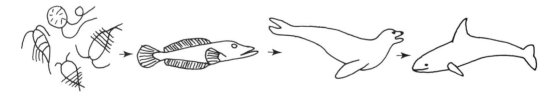

Fig. 5-5. The marine food chain begins with simple microorganisms, which are eaten by progressively larger species.

So much garbage is generated in the big cities that the only place left to put it is into the sea. For a typical ocean dumping site along the East Coast, barge loads of waste are taken out to about 100 miles and dumped beyond the continental shelf (FIG. 5-6). After a

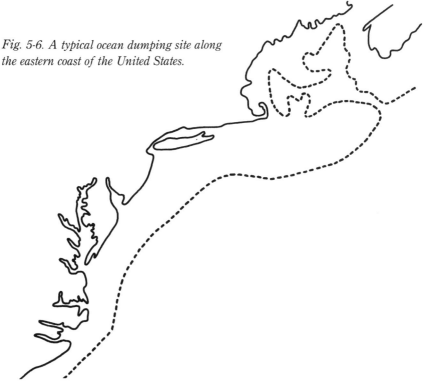

Fig. 5-6. A typical ocean dumping site along the eastern coast of the United States.

day or so, the pollutants are diluted by a factor of about 50,000:1. However, it is still questionable whether even a dumping dilution of 100,000:1 is adequate. The pollutants concentrate in regions where the seawater density changes, such as thermoclines and ocean fronts. Unfortunately, these frontal areas are also the feeding grounds for fish.

Of all the alternative disposal methods for hazardous waste, none has engendered as much controversy as the incineration of toxic substances. One plan envisions using an incinerator ship to burn toxic wastes offshore. It would be a stopgap measure to treat toxic liquids until better methods to reduce or recycle waste on land are developed. Ocean incineration would only be suitable for treating five to eight percent of all hazardous waste. However, the chemicals that could be destroyed by the technology are among the most toxic.

Incineration at sea is one of the few methods available to detoxify hazardous wastes that are highly chlorinated. However, many questions concerning the potential risks to health and the environment are unresolved. Better methods need to be developed to measure whether dangerous compounds are actually destroyed by incineration and to identify what compounds are being emitted into the atmosphere—ultimately ending up in the ocean.

Many thousands of water wells across the nation have been contaminated by highly concentrated chemicals that spread through aquifers and exceed the Federal safe drinking water limits. In California's Silicon Valley, solvents used in the manufacture of computer chips have leaked from buried storage tanks into the water supplies of several communities. A New Jersey landfill that contained nine million gallons of hazardous wastes contaminated the aquifer and forced the shutdown of nearby water wells in Atlantic City.

In Florida alone, 6000 lagoons and ponds are filled with toxic waste, and contaminants were found in the groundwater, forcing the shutdown of more than 1000 wells. It is estimated that $500 billion would be required to clean up the nation's toxic sites. The dumping of toxic wastes into the ocean might irreversibly change aquatic ecosystems. The ever-mounting stockpiles of nuclear waste are crucial because they remain a hazard to life for thousands of years.

NUCLEAR WASTE DISPOSAL

The disposal of radioactive wastes from nuclear power plants and nuclear weapons manufacturing (FIG. 5-7) has received much attention in recent years because of increasing concerns over the long-term environmental effects and because of the expansion of nuclear technology throughout the world. High-level nuclear wastes are the most difficult

Fig. 5-7. The Rocky Flats nuclear components plant near Golden, Colorado.

radioactive waste materials to dispose of because of their high radiation, heat output, and longevity. Some substances, such as plutonium, require millions of years to decay.

It is generally thought that the best place to store nuclear waste is underground (FIG. 5-8), and much effort has been focused on exploring for stable geologic formations.

Fig. 5-8. An underground nuclear waste disposal in a salt bed 2000 feet below the surface, near Carlsbad, New Mexico.

Salt domes and granite are the most stable geologic formations on the continents. However, mine repositories are very expensive and require additional costs for backfilling and shaft sealing. Also, we have no absolute guarantee that the nearby groundwater systems will not be contaminated after the containers age and begin to leak.

Therefore, scientists must make reliable predictions concerning the possible migration of radioactive fluids through geologic formations surrounding the repository. The formation must remain stable for one million years or so without earthquakes or other geologic activity. It must also be guarded against intrusion and theft for countless generations to come.

The United States has been looking for a permanent repository of nuclear wastes, which are presently stored at nuclear power plants and other dump sites. Some of these sites have been leaking; the consequential danger is of contaminating rivers and groundwater systems. The underground site must be equipped to hold up to 100,000 tons of nuclear wastes, which are sealed in protective casks (FIG. 5-9). The below-ground storage facility must also be sealed against the contamination of the groundwater system and against human intrusion for 10,000 years or more.

Fig. 5-9. Cutaway view of nuclear waste containers at the underground nuclear disposal site near Carlsbad, New Mexico.

The present design calls for a four-square-mile chamber mined out of solid rock or salt 1000 to 4000 feet underground. The total cost for the repositories would range from 25 billion to 30 billion dollars, which would ultimately have to be passed on to consumers through higher electric bills.

Transporting the nuclear wastes to dump sites in the West is also hazardous—especially since 85 percent of the waste is scattered in various parts of the nation, but mostly in the East. According to Department of Energy estimates, it would take 17 truck loads per day for the next 20 years to move all the wastes to the burial sites. With that many vehicles carrying nuclear wastes on American roads at any given time, the prospect of an accident and a consequent spill is far too great for communities along the routes to accept. Packaging the nuclear wastes in solid form or designing containers that are virtually indestructible might alleviate the danger from road accidents.

It has even been suggested that the nuclear wastes should be placed in wells drilled deep into the seabed. The idea is that certain parts of the ocean floor are the most stable environments on Earth, whereas the land is always subject to earthquakes, mountain building, and erosion. Once the nuclear waste containers are sealed off against the sea, the constant rain of detrital material washed off the continents will continue to bury them under thick layers of sediment. Areas containing natural resources, such as fisheries, petroleum reserves, or mineral deposits should be avoided for fear of disturbing the burial site and contaminating the ocean.

As human populations continue to grow and the demand for nuclear-generated electricity continues to rise (presently accounting for about 15 percent of the world's total generating capacity), a viable solution for the storage of nuclear wastes must soon be found if our living areas are to remain free of man-made radioactive poisons.

6

Dwindling Resources

WE have been blessed with a world that is rich in natural resources. The exploitation of mineral and energy resources has greatly improved people's lives. An unfortunate byproduct of this advancement is pollution and the health hazards that it entails. Generally, however, most industrial nations have prospered in a manner that had been unprecedented in man's long climb up the ladder of progress. Unfortunately, the depletion of natural resources could threaten future advancement. We are depleting fossil fuels at a rate that is 100,000 times faster than they are formed. Only through the conservation of natural resources and the exploration of alternative energy sources can the wealth of the Earth be preserved for future generations.

BLACK GOLD

Petroleum presently supplies about 50 percent of the world's energy needs (FIG. 6-1). Soon after the turn of the century, however, the supply of oil might begin to fail to meet the increasing demand of growing industry. World energy consumption is expected to increase over 50 percent by the year 2010. Of the over one trillion barrels of oil thus far discovered, 30 percent or more has already been consumed. Presently, the world uses 65 million barrels of oil per day and the United States consumes 17 million barrels per day—about 25 percent of the total. In Europe and Japan, an average person uses between 10 and 30 barrels of oil annually, whereas an American consumes over 40 barrels per year.

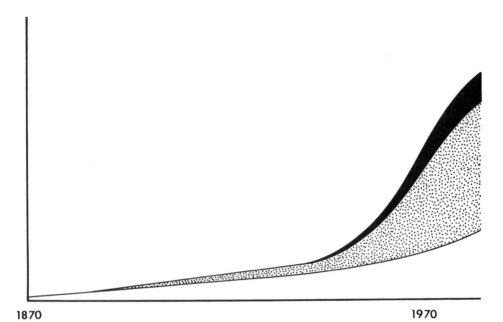

1870   1970

Fig. 6-1. The rate of increase in energy use. White indicates coal use, stippled indicates oil and gas use, and black indicates electricity use.

As oil production levels off and then falls, alternative fuels will have to be developed to meet the demand for energy, which will continue to grow—even with conservation methods in place. For oil importing countries, which commandeer about 50 percent of the oil on the market, this will require a transition from a dependence on oil to a greater reliance on other fossil fuels, nuclear energy, and renewable energy sources.

Over the last two decades, offshore drilling for oil and natural gas in shallow coastal waters (FIG. 6-2) has become extremely profitable. About 20 percent of the world's oil and about 5 percent of the natural gas production is offshore. Projections indicate that in the future, perhaps twice as much oil will be pumped from the seas than from the land. Unfortunately, a great deal of offshore oil leaks into the oceans, amounting to as much as 2 million tons per year. This spillage could create an enormous environmental problem as oil production continues to increase.

Interest in offshore oil began in the mid-1960s and drilling stepped up a decade later following the 1973 Arab oil embargo, when American motorists were forced to wait in long lines at gas stations. Important finds, such as Prudhoe Bay on Alaska's North Slope (FIG. 6-3) and the North Sea off of Great Britain, resulted from intensive exploration for new reserves of offshore oil. The nation's desire for energy independence prompted oil companies to explore for oil in the deep oceans. Many difficulties were encountered, however, including storms at sea and the loss of personnel and equipment, which could not justify the few discoveries that were made.

In the early 1980s, the Department of the Interior estimated the amount of remaining offshore oil and natural gas reserves that were large enough to be commercially exploited around the United States. By the mid-1980s, the Department cut its original estimates of oil reserves in offshore fields by 50 percent. The revised figures reflected

Fig. 6-2. A semisubmersible drilling rig in the Mid-Atlantic outer continental shelf.

Fig. 6-3. An oil tanker approaches a berth at the Valdez terminal of the transAlaska pipeline, which carries crude oil from Alaska's North Slope.

Black Gold 67

the fact that oil companies were virtually empty-handed after four years of exploratory drilling in the highly promising areas of the Atlantic and off the coast of Alaska.

It was hoped that these regions would help keep the nation well supplied with oil. Without reliable reserves, the United States could become dangerously dependent on foreign sources for meeting its demand. This prospect became clear when Iraq invaded Kuwait in August 1990—even though the Persian Gulf region only supplies 7 percent of the world's oil. The United States responded to this threat with the greatest military deployment since World World II. The resulting Persian Gulf War in January 1991 underlines how dependent western economies have become on imported oil. The United States imports about 85 percent of its petroleum; the economy could become disastrous if even a fraction of this oil is cut off.

It is doubtful that the consumption of fossil fuels will decline in the industrial nations over the next 30 years; it is more likely to increase. Developing countries will want to industrialize to improve their standard of living. Furthermore, it is estimated that by the year 2035, most of the petroleum reserves will be depleted. Unless some other safe alternatives, such as fission, fusion, solar, and geothermal energy, are developed and rapidly exploited, industrial plants will have to convert to coal, whose resources are barely touched, but whose environmental impacts are much more severe.

DIRTY COAL

Coal presently meets about 25 percent of the world's energy needs. Its peak usage was in the 1920s, when it accounted for more than 70 percent of fuel use. The United States has increased its coal consumption by about 70 percent since the decade of the 1970s. The coal is used mainly for electrical power generation in huge coal-fired plants. The generation of electricity accounts for about 75 percent of all coal consumed in the U.S.

Because much of the coal from Eastern underground mines has a high sulfur content, utilities are forced to burn Western coal, which is mined from massive open pits (FIG. 6-4). In order to keep up with demand, the United States will have to mine about 60 percent more coal by the turn of the century. The total world coal production is about 5 billion tons annually. The United States accounts for about 50 percent of the coal mined and consumed by the free world.

Globally, coal reserves far exceed all other fossil fuels combined. They are sufficient to support large increases in consumption well into the next century. The amount of economically recoverable coal is nearly one trillion tons; at the present rate of consumption, these reserves could last well over 200 years. The United States possesses about 50 percent of the free world's economic coal reserves (FIG. 6-5), which are practically untouched. China and the Soviet Union contain most of the rest.

Because coal is the cheapest and most abundant fuel, it will be the most favorable alternative source of energy to replace petroleum when it runs out. Unfortunately, coal combustion yields significantly more carbon dioxide per unit of heat than oil and natural gas and it contributes substantially to greenhouse warming.

If consumption continues at its present furious pace, all the nation's economically recoverable coal might go up in smoke in only a few decades. This consumption could tremendously overload the atmosphere with carbon dioxide and other dangerous chemi-

Fig. 6-4. Open-pit coal mining at the West Decker mine in Montana.

Fig. 6-5. Location of coal deposits in the United States.

cals. Another problem with burning massive quantities of coal and petroleum is that millions of tons of sulfur dioxide and nitrogen oxides are discharged into the atmosphere each year. These gases, when combined with oxygen and atmospheric water vapor, produce acid rain. They also rise to the upper atmosphere and destroy ozone.

MINERAL EXPLOITATION

The insatiable appetite for petroleum and mineral ores to maintain a high standard of living in the industrialized world as well as to improve the standard of living in developing countries might lead to the depletion of known oil and high-grade ore reserves by the middle of the next century (TABLE 6-1). After that, low-grade deposits will have to be

COMMODITY	RESERVES*	CONSUMPTION RATE IN YEARS RESOURCES
Aluminum	250	800
Coal	200	3000
Platinum	225	400
Cobalt	100	400
Molybdenum	65	250
Nickel	65	160
Copper	40	270
Petroleum	35	80

TABLE 6-1. The Future of Some Natural Resources.

*Reserves are recoverable resources with today's technology

worked, dramatically increasing the cost of goods and commodities. However, we have barely scratched the surface in the search for minerals. Immense resources lie at great depth, awaiting the technology to bring them to the surface.

Large quantities of oil and gas exist in the Middle East, the Gulf Coastal region, the Rockies, the North Slope of Alaska, and the North Sea. Huge untapped reserves of oil exist in oil-shale deposits in the western United States (FIG. 6-6) with a potential oil content that could exceed all other petroleum resources in the world. Abundant coal deposits exist in the western United States, Canada, South Africa, and Asia.

Iron is the fourth most abundant element in the earth's crust and it is found in economical deposits on all continents. However, ore grades generally must exceed 30 percent to be mined profitably. The Mesabi Range of northeast Minnesota is the major supplier of iron ore for the United States. The iron is contained in a banded iron formation that was laid over two billion years ago. The Clinton iron formation is the chief iron producer in the Appalachian region, from New York to Alabama.

IN-SITU RETORTING-OIL SHALE

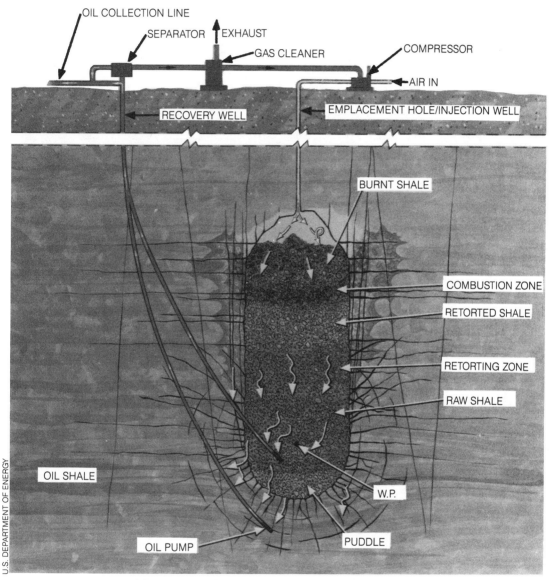

OIL COLLECTION LINE

SEPARATOR ↑ EXHAUST

GAS CLEANER

COMPRESSOR

AIR IN

RECOVERY WELL

EMPLACEMENT HOLE/INJECTION WELL

BURNT SHALE

COMBUSTION ZONE

RETORTED SHALE

RETORTING ZONE

RAW SHALE

OIL SHALE

W.P.

OIL PUMP

PUDDLE

Fig. 6-6. An artist's concept of in-situ retorting of oil shale.

Important deposits of copper, lead, zinc, silver, and gold are found in the Rocky and Andes Mountain regions. A variety of other metallic deposits exist in the mountains of southern Europe and in the mountain ranges of southern Asia. In the western United States, important reserves of uranium spurred the great uranium boom of the mid-20th century. Important reserves of phosphate, used for fertilizers, were found in Idaho and

Fig. 6-6. Continued.

SURFACE RECOVERY PLANT

MULTIPLE ARRAY-PLAN VIEW

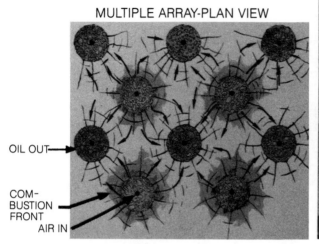

OIL OUT →

COM-
BUSTION
FRONT
AIR IN

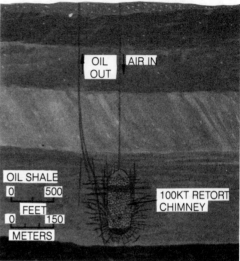

adjacent states. Nonmetallic minerals, such as sand, gravel, clay, salt, limestone, gypsum, and phosphates, are mined in great quantities over the world.

Gold is mined on every continent, except Antarctica. In Africa, the best gold deposits are in rocks as old as 3.4 billion years. In North America, the best gold mines are in the Great Slave region of northwest Canada, which has over one thousand known deposits. The Mother Lode in California and other Western states was responsible for the gold rush days of the mid-19th century (FIG. 6-7). The Comstock Lode in Nevada was responsible for one of the largest mining booms in the history of the opening of the American West. The rich silver and gold mines in South America were largely responsible for the Spanish conquest of the Inca Empire.

CONSERVING RESOURCES

Twenty percent of the human population consumes more than 70 percent of the world's resources. The United States, with only 5 percent of the world's population, consumes 25 percent of the resources. The few rich nations are more responsible for the pollution and degradation of the environment than all the rest of the world combined. The U.S. per capita consumption of resources and production of pollution is roughly four times greater than other industrialized nations.

It is no wonder that developing countries envy our lifestyle and are rushing to develop their own resources and participate in the prosperity. Unfortunately, this additional burden of atmospheric pollution and carbon dioxide might worsen the greenhouse problem. Increased efficiency of energy use and alternative fuels might help developing countries raise their standard of living without significantly increasing energy use or pollution.

The long-term increase in atmospheric carbon dioxide, as much as 25 percent since 1860, is the result of an accelerated release of carbon dioxide by the combustion of fossil fuels. The present consumption of fossil fuels yields on average about 1.1 tons of atmo-

Fig. 6-7. A miner's camp on King Solomon Mountain, Colorado in 1875.

spheric carbon for each of the world's 5.4 billion people every year. Americans release nearly six tons per person per year, which amounts to 1.2 billion tons (about 25 percent of the total).

Large-scale conservation measures might be required to preserve plant and animal species that are threatened by global climate change—especially if it occurs too rapidly. The two response strategies for combating climatic change are *adaptation*, involving such measures as moving to a cooler climate or building coastal defenses against a rising sea, and by *limitation*, which directly involves limiting or reducing the emissions of greenhouse gases. Perhaps the most prudent response to climatic change would utilize both of these measures.

Conservation can curtail the effects of global warming and it will result in large part from the consequences of improved energy efficiency and the development of nonpolluting substitute energy sources. However, more conservation can only attack a portion of the carbon dioxide problem, not solve it. The world also needs a constitution for the atmosphere that is similar to the Law of the Sea, because one nation's pollution affects all nations.

FUTURE ENERGY

Industrial nations will face a crisis in the upcoming years if alternative sources of energy are not found before fossil fuels start to run out. Nuclear energy was once thought to be the answer to the world's energy problems. Early promoters painted an overly optimistic future, saying that electricity generated by nuclear power would be too cheap to meter. However, heavy government regulations, frequent delays, and accidents (FIG. 6-8) have skyrocketed the cost of nuclear power plants in recent years.

Fig. 6-8. The Three Mile Island nuclear power plant, Harrisburg, Pennsylvania. It underwent a reactor accident in 1979, which necessitated reassessment of nuclear safety in the United States.

Therefore, it seems that nuclear energy will play only a minor role in meeting this nation's electrical needs. However, in order to combat atmospheric pollution and the greenhouse effect, a reassessment of nuclear energy might be required because nuclear generating plants are essentially nonpolluting. The safety of the plants will have to be ensured and nuclear wastes will have to be managed properly before nuclear energy can be considered a viable alternative. Many European countries, particularly France, rely heavily on nuclear energy to replace costly fossil fuels.

Unfortunately, the nuclear accident at the Chernobyl power station near Kiev in the Soviet Union in April 1986 set world opinion against nuclear electrical power plants, spurring the search for alternate energy sources. One such alternate source is *solar energy* (FIG. 6-9). The Sun itself is a nuclear reactor that radiates a tremendous amount of energy. The sunlight reaching the Earth is only about one billionth of the total amount of solar energy that is radiated into space. However, the amount that reaches the Earth in one year is equivalent to about 15,000 times the world's present energy supply. On the Earth, an area of one square meter (about 10 square feet) of solar energy is equivalent to 600 watts of electricity.

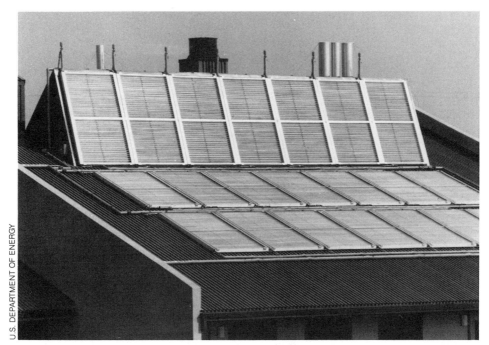

Fig. 6-9. Solar panels on a laboratory at the University of California, Davis, California.

The direct conversion of sunlight into electricity is accomplished by the use of *photovoltaic* (solar) *cells*, which have an optimum efficiency of about 20 percent. The manufacture of solar cells is very expensive, however, making their use uneconomical on a large scale. But less efficient solar cells might be manufactured in mass quantities at a greatly reduced price, provided that the demand is high enough. Promoters claim that home solar electrical plants can offer substantial savings over the cost of running power lines to rural homes.

Another method of converting sunlight into electricity is by focusing it into a powerful narrow beam, using banks of *heliostatic mirrors*, which automatically track the Sun as it travels across the sky. The light beam is focused onto a central receiving station, where the intensified light heats a boiler and the superheated steam drives a steam turbine generator.

One obvious drawback of solar generating plants is that their efficiency is poor on cloudy days, and they do not operate at night. Therefore, to take full advantage of solar energy, the best places for installing solar power plants is in areas where the Sun shines for long periods throughout most of the year, such as in the southwest American desert. This location is also ideal because solar power plants require large tracts of land, which is relatively inexpensive in this region. Solar power stations still cannot compete economically with conventional fossil fuel generating stations. This situation might change, however, when fossil fuels become scarce and expensive.

Commercial and residential buildings can utilize solar energy to heat water, which is used to supplement conventional water heaters and furnaces. These systems can provide substantial savings on utility bills, while conserving unrenewable energy resources. The sunbelt states, which receive a generous supply of sunlight, can take advantage of this type of solar energy. The systems promise to pay for themselves in utilities savings in about a decade.

Air and water currents are other forms of solar energy. In windy localities, such as seacoasts, where the offshore and onshore wind currents are fairly reliable, windmills can be used to generate electricity (FIG. 6-10). The wind also drives ocean waves, which

Fig. 6-10. A large vertical axis wind turbine that is located near Bushland, Texas.

can be harnessed to produce electricity. Trapping tidewaters in enclosed bays is another means of generating electricity from the power of falling water.

The most successful use of solar energy is *hydroelectric power*. Because hydroelectric projects are expensive and the most accessible sites have already been utilized, hydroelectric power will probably not figure significantly in future energy increases. One little-known energy source is called *ocean thermal-energy conversion* (OTEC), which takes advantage of the temperature difference between thermal layers of the ocean to generate electricity with a large, low-pressure steam generator.

Like solar energy, the potential for geothermal energy is enormous. The Earth's interior is a natural nuclear power reactor, in which heat is generated by the decay of radioactive elements. Many steam and geyser fields around the world are associated with active volcanism, which makes these areas ideal for tapping geothermal energy for steam heat and electrical power generation (FIG. 6-11).

Fig. 6-11. A geothermal generating plant at the Geysers, San Francisco, California.

An energy source that is both renewable and essentially nonpolluting is *fusion*. The fuel used for fusion is deuterium, which is abundant in seawater. The energy from the fusion of deuterium in a tank of seawater the size of an olympic swimming pool could provide the electrical needs for a city with 250 thousand people for an entire year. Fusion is also safe; its byproducts are energy and helium, a harmless gas that escapes into space. It remains to be seen if the development of fusion is timely enough to solve the many energy problems that plague the world of the future.

7

Burgeoning Populations

TWO million years ago, our ancient ancestors numbered perhaps only 100,000 people. When agriculture was invented, roughly 10,000 years ago, from 5 to 10 million people were in the world. During the first dynasty in Egypt, around 5000 years ago, the population rose to 100 million. At the height of the Roman Empire, during the first century A.D., the world's population more than doubled. It doubled again at the beginning of the Industrial Revolution and the first billion mark was reached around 1800.

Now less than two centuries later, the number of people in the world has increased more than fivefold and it is expected to grow to 10 billion by 2030. In effect, a long period of slow growth was followed by a brief period of rapid growth. Only in the past few decades have humans brought about major global changes comparable in magnitude to those wrought by nature over long periods of geologic time. The damage to the Earth will continue to escalate as long as burgeoning human populations continue to grow out of control.

THE AGRICULTURAL REVOLUTION

Even at its earliest onset, farming was so productive that it could support many times more people in a given area of land as hunting and gathering. It is therefore not surprising that agriculture not only supported, but encouraged population growth. Farming was labor intensive, it required large families to till and harvest the land. Increasing populations also required an intensification of food production, often with disastrous consequences to the land.

The advent of agriculture brought with it sedentism, and permanent villages sprang up. The ownership of land made people territorial and they fought furiously to maintain control. Increasing populations required innovations to increase food production, such as improved farming techniques, which ultimately led to irrigation and the invention of the plow. These actions were often not enough, however, and overpopulation forced people to migrate from the region.

About 6000 B.C., agriculture spread to northern Europe and Great Britain. At this time, Europe was largely forested and huge tracts of land were cleared for planting crops and grazing cattle. It is even suspected that the felling of so many trees might have significantly altered the European climate. The expansion of agriculture into areas with the most easily cultivable soils resulted in overpopulation, forcing people to move to less desirable locations.

About 4000 years ago, the climate grew drier and the present deserts began to form (FIG. 7-1). The plains of Mesopotamia in southwest Asia had abundant fertile soil, but no

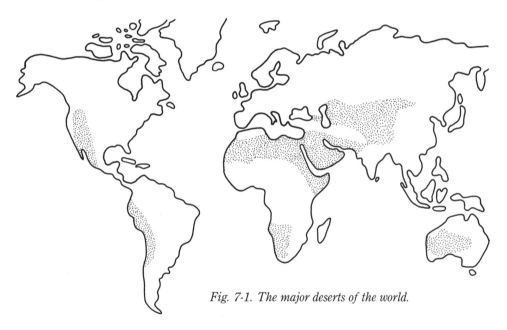

Fig. 7-1. The major deserts of the world.

water for the crops. Large irrigation projects were erected, which required the hard labor of hundreds of thousands of people and a system of centralized authority to rule over them. In merely one thousand years (by about 3000 B.C.), a loose-knit egalitarian society was transformed through agriculture into a strongly ruled authoritarian society that was equipped with masters and slaves.

Population growth, which previously had been held in check by the harsh demands of nature, soared after the invention of agriculture. In a mere 10,000 years following the last ice age, the number of humans increased geometrically, from about five million to one thousand times as many today. As many people live now as ever had since the dawn of man. A deadly combination of drought, disease, and infestation could conspire against humanity, possibly causing the greatest famine and the largest death toll from starvation that the world has ever known.

SOIL LOSS

One of the greatest limiting factors to increasing human populations is the degradation of the land by soil erosion (FIG. 7-2). Every year, the world's farmers are trying to

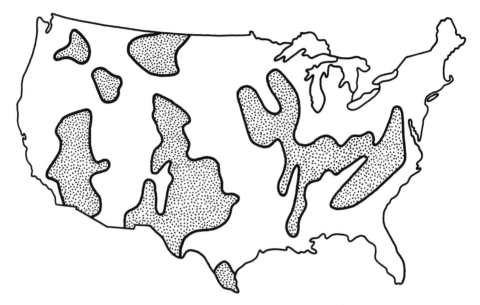

Fig. 7-2. Areas in the United States that are affected by soil erosion.

feed 100 million more people with 25 billion fewer tons of topsoil. Prior to the advent of agriculture, natural soil erosion rates were about 10 billion tons per year, slow enough to be replaced by the generation of new soil. Estimates of present soil erosion rates, however, are two to three times greater. Therefore, we are losing soil faster than nature is putting it back. As much as 30 percent of the world's cropland is losing soil at a rate that is undermining any long-term agricultural productivity. World food production per capita will eventually fall off if the loss of topsoil continues at its present excessive rate.

Erosion rates depend on the amount of precipitation, the topography, the type of rock and soil (TABLE 7-1), and the amount of vegetative cover. Short-term efforts to increase worldwide crop production through deforestation, wetlands drainage, irrigation, artificial fertilization, genetic engineering, and improved farming techniques will ultimately be defeated if the topsoil is washed away.

In order to keep up with an ever-growing human population, which adds another 100 million mouths to feed every year, farmers are abandoning sound soil conservation practices and are implementing more intensified farming methods. These unsound methods include less rotation of crops, greater reliance on row crops, more plantings between fal-

CLIMATE	TEMPERATE (HUMID) > 160 IN. RAINFALL	TEMPERATE (DRY) < 160 IN. RAINFALL	TROPICAL (HEAVY RAINFALL)	ARCTIC OR DESERT
Vegetation	Forest	Grass and brush	Grass and trees	Almost none, no humus development
Typical area	Eastern U.S.	Western U.S.		
Soil type	Pedalfer	Pedocal	Laterite	
Topsoil	Sandy, light colored; acid	Enriched in calcite; white color	Enriched in iron and aluminum, brick red color	No real soil forms because no organic material. Chemical weathering very low
Subsoil	Enriched in aluminum, iron, and clay; brown color	Enriched in calcite; white color	All other elements removed by leaching	
Remarks	Extreme development in conifer forest abundant humus makes groundwater acid. Soil light gray due to lack of iron	Caliche - name applied to accumulation of calcite	Apparently bacteria destroy humus, no acid available to remove iron	

TABLE 7-1. Summary of Soil Types.

low periods, and extensive use of chemical fertilizers, rather than natural organic fertilizers, which help keep the soil on the farm.

Throughout the world, most of the arable land is already under cultivation, and efforts to cultivate substandard soil leads to poor productivity and abandonment, resulting in severe soil erosion (FIG. 7-3). Marginal lands, which are often hilly, dry, or contain only thin, fragile topsoils, and therefore erode easily, are also forced into production. As world populations continue to grow geometrically on a planet whose resources are dwindling rapidly, the vast majority of people will be forced to live barely at subsistence. As a hungry world keeps crowding onto worn-out soils, man-made deserts will continue to spread across large parts of the Earth.

MAN-MADE DESERTS

The Fertile Crescent, between the Tigris and Euphrates Rivers in what is now Syria and Iraq, once fed as many as 25 million people. Today, however, it is infertile from over-irrigation and salt accumulation in the soil by Sumerian farmers 6000 years ago. Over one thousand years ago, nomads from the Sahel region of central Africa lived by hunting and

Fig. 7-3. Severe gully erosion on a pasture in Shelby County, Tennessee, as a result of cultivating unsuitable soil that should have been left growing grass.

herding. The Sahel, which lies to the south of the Sahara desert, is about 250 miles wide and it extends west to east across central Africa (FIG. 7-4). The Sahel was once mostly a tropical forest. But the nomads cut and burned the trees to improve grazing in the region, thus turning a natural forest into a grassland.

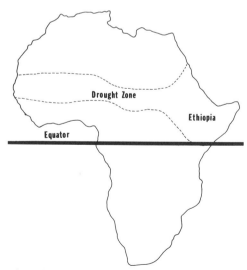

Fig. 7-4. The Sahel region of central Africa is a broad zone of continuous drought that was caused mainly by human activities.

Last century, as foreign powers carved up Africa amongst themselves, the people of the Sahel were forced to remain in the region as farmers and herdsmen. As a result of overgrazing, the soil was weakened, creating an extensive man-made desert. The Sahel is now being overrun by the sands of the Sahara desert, which are steadily advancing, engulfing everything in their path. The process is exacerbated because, without vegetation, the land is subjected to flash floods, higher evaporation rates, and tremendous dust storms.

The denuded land also has a higher *albedo* (TABLE 7-2), which decreases rainfall and denudes more land. Thus, man-made deserts march across once fertile acres. Throughout the world, perhaps as much as 30 to 50 percent of what was once arable land is now rendered useless by erosion and desertification (FIG. 7-5). Moreover, from 50 percent to 75 percent of all irrigated land will be destroyed by salt accumulation in the soil by the end of this century.

TABLE 7-2. Albedo of Various Surfaces.

SURFACE	PERCENT REFLECTED
Clouds, stratus	
< 500 feet thick	25-63
500-1000 feet thick	45-75
1000-2000 feet thick	59-84
Average all types and thicknesses	50-55
Snow, fresh-fallen	80-90
Snow, old	45-70
White sand	30-60
Light soil (or desert)	25-30
Concrete	17-27
Plowed field, moist	14-17
Crops, green	5-25
Meadows, green	5-10
Forests, green	5-10
Dark soil	5-15
Road, blacktop	5-10
Water, depending upon sun angle	5-60

The Persian Gulf oil spill off the coast of Saudi Arabia.

Exxon workers clean up beach on the coast of Prince William Sound, Alaska soiled by the Exxon Valdez *oil spill.*

A landfill in Raleigh, North Carolina.

Lake Santa Margarita in California down to 6 percent capacity due to a long period of drought.

The remains of an ancient burned forest.

Windmill technology development at the Rocky Flats Plant near Golden, Colorado.

Forest fires contribute to air pollution by emitting carbon dioxide, methane, and other gases into the atmosphere.

Primary sources of industrial pollution include power plants, smelters, and refineries that emit oxides of sulfur and nitrogen along with other gases into the atmosphere.

Fig. 7-5. Soil drifting from an unprotected corner of a wheat field in Chase County, Nebraska.

One hundred years ago, global rain forests covered an area of about twice the size of Europe. Now that area has been cut in half to make room for new cropland. As developing nations attempt to raise their standard of living, one of the first things they do is clear the forests and wetlands for agriculture. Much of the land is cleared by slash-and-burn methods, whereby trees are set ablaze and their ashes used to fertilize the thin, nutrient-poor soil.

After a year or two of improper farming or grazing methods, the soil is worn out and farmers are forced to move on. The abandoned farms are then subjected to severe soil erosion because no plant cover remains to protect against the effects of wind and rain. Once the forests are leveled, the ground is often laid bare to the elements and the soil severely erodes. When the soil disappears, rain forests that have been in existence for as long as 30 million years disappear forever.

The process of desertification severely degrades the environment and is caused mainly by human activity and climate. The loss of topsoil takes millions of acres of once fertile cropland and pastureland out of production each year. After the land is denuded, only the coarse sands remain (FIG. 7-6) and a man-made desert is created. The problem is exacerbated when the land is subjected to flash floods, high erosion rates, and dust storms, which sweep the sands from place to place.

Desertification is occurring all over the world, but it is most prevalent in central Africa, where the sands of the Sahara desert march across the Sahel region to the south. The process of desertification is also self perpetuating because the light-colored sands reflect more sunlight, creating permanent high-pressure regions that block out weather

Fig. 7-6. The soil profile, showing dark, organic-rich topsoil and sandy, infertile subsoil below. Measurements are marked in feet.

systems and lessen rainfall. The land is also subjected to flash floods and dust storms, which transport the soil out of the region.

Desertification is also occurring in the tropical rain forests, which are being cleared on an unprecedented scale for cattle grazing. After a couple of years of extensive agriculture, the soil is depleted of its nutrients, and because most of the world's farmers cannot afford expensive fertilizers, the infertile land is abandoned. Meanwhile, heavy downpours wash away the denuded topsoil, often leaving bare bedrock behind.

As a large part of the rain forest is being destroyed, precipitation patterns within the forested areas are changing, which could possibly turn them into a man-made desert. The denuding of the world's forests increases the Earth's albedo, and causes a loss of precipitation. This problem strains the forests more and subjects them to infestation and disease, causing an additional die-out of trees.

It also appears that the removal of the rain forests is having a dramatic effect on the global climate, resulting in lesser amounts of rainfall. High evaporation and transpiration rates within the forests themselves help to create more clouds, which contribute to the prodigious rainfall that these areas receive. When the forests are removed, however, the cycle is broken and the torrential rains now only cause severe flash floods.

THE OVERCROWDED PLANET

World populations are growing so explosively and modifying the environment so extensively, that we are inflicting a global impact of unprecedented dimensions. Humans presently consume about 40 percent of the terrestrial net primary production, which is the total amount of energy stored by green vegetation on the land surface. The level of human growth over this century alone has been staggering. At the present rate of growth, the population is expected to reach the 10 billion mark by 2030. Then, humans will require twice the consumption of today's world net primary production. This notion is preposterous considering the destructive impacts of today's level of human activities.

Rapid population growth has already stretched the resources of the world and the prospect of future increases raises serious doubts as to whether the planet can continue to support people's growing requirements. Up to a tenfold increase in world economic activity over the next 50 years would be required to keep up with basic human needs. The biosphere cannot possibly tolerate this situation without irreversible damage.

The most viable solution to this dilemma is to limit population growth (FIG. 7-7). The relationship between population density and food supply is obvious. In the natural world,

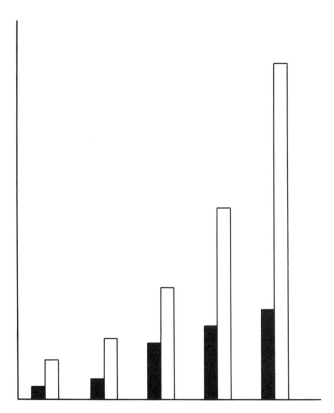

Fig. 7-7. Relative population growth since 1750 in developed countries (black bar) and in undeveloped countries (white bar).

animal populations are limited by the amount of food available, but human beings do not follow this example. People can and do survive on less than the optimum amount of food, although they do not live very long at this level of bare subsistence.

According to the 19th century English economist Thomas Malthus, this food supply is just sufficient to sustain life obtained by the maximum work effort that the population can exert. Malthus believed that human populations are limited by the food supply because they grow geometrically, whereas food production only increases arithmetically. He also indicated that humans increase their numbers beyond their means of subsistence until famine, war, and disease wipe out the excess.

Except for extreme shortages of food, most adults do not die from an inadequate diet. Eventually, however, vitality, health, and ability to work suffers. Also, damage to the intestinal tract, caused by parasites and infection, makes the body less efficient, so a higher caloric intake is required to stay alive. But the effects on young children who survive mortality are often irreversible, leading to disease, stunted growth, blindness, and mental retardation.

The undeveloped countries are no longer capable of producing much (if any) food surpluses in good years, which could provide reserves for lean years. As a result, they are extremely vulnerable to climatic fluctuations. Before World War II, the undeveloped countries as a whole were exporters of cereals. Since then, they have become importers, with much of the tonnage obtained under food aid programs. Also, when food prices rise as a result of a bad harvest, poor families must forego other needs, further lowering their standard of living in order to obtain food.

In the early 1970s, an internationally managed world food bank, which could augment supplies of food during years of drought, pestilence, or disease was suggested. Unfortunately, the 1973 Arab oil embargo undermined such efforts by raising the cost of pumping groundwater for irrigation (FIG. 7-8) and by increasing the price of fertilizer, which was already in short supply. The developed nations, with highly mechanized farming, were particularly hard hit with the inflated oil prices, causing higher food prices that the poorer countries could not afford.

Fig. 7-8. A center-pivot irrigation system on a Nebraskan farm.

A WORLD OF HUNGER

Many scientists who study human populations believe that the world is already at carrying capacity, which is the ability of the land to support people. A population exceeds carrying capacity when it cannot be maintained without rapidly depleting nonrenewable resources. By that definition, the world is already vastly overpopulated. It will be difficult to feed, clothe, shelter, and employ additional people at more than subsistence. Over the past two decades, at least 200 million people have died of hunger-related diseases. Rapid population growth has stretched the resources of the world; the prospect of future increases raises serious doubts as to whether the planet can continue to support people's growing needs.

In developing countries that have the highest birth rates, economic growth has come to a standstill. The few economic gains that are made are quickly eaten. The developing nations of Asia, Africa, and South America are in a desperate race to keep food supplies up with population growth. Sub-Saharan Africa, which comprises 47 countries, has a population of nearly 500 million people, and that number is expected to double in 20 years. Already, 100 million Africans are malnourished, and 25 percent of the children die before the age of five. When droughts occur and famine strikes in these regions, people are placed in grave danger—especially if, for political or other reasons, food imports cannot meet the demand, the prospect of mass starvation increases.

Ninety percent of all the world's food is from only about one dozen crops, and most strains are genetically undiversified. Disease and infestation targeted at these specific strains could wipe out a nation's entire harvest. When agriculture can no longer supply the necessary food, people suffer from famine. During favorable climates, populations tend to grow well beyond the limits imposed by unfavorable climates, when harvests are poor. Mass starvation might result from severely reduced crop yields as a result of drought, infestation, disease, frost, or storm damage (FIG. 7-9).

The leading food exporters already have most of their arable lands in production. Most of the production, however, is consumed or sold to other developed nations at prices that the developing countries cannot afford. During the decade of the 1970s, American farmers put into production an additional 60 million acres, an area larger than the state of Kansas, in order to help feed a hungry world. Much of this increased land was substandard, including sloping, marginal, and fragile soils that erode easily. Under increasing pressure for more food production, normally fallow fields are being cultivated, which quickly wears out the soil.

The total quantity of food directly and indirectly consumed by the human population is staggering; it amounts to roughly one ton per person per year (about 5 billion tons annually). Nearly 50 percent of the total crop tonnage and 75 percent of the energy and protein content is supplied by cereal grains (FIG. 7-10). A large fraction of these grains is eaten by domestic animals, which consume 4 to 7 calories of grain for every calorie of meat they produce.

The average individual food intake is 2700 calories per day for developed nations and 1800 calories per day for developing countries. However, the diets of the poorer countries are not nearly as nutritious as those of the richer nations. For the poorest 20 percent of the world's population, their diet, which consists mostly of cereal grains, tubers, and other starchy roots, falls below the body's requirement for a normally active, healthy life.

Fig. 7-9. Severe hail damage to a cornfield.

Fig. 7-10. A large grain silo in Pierceville, Kansas.

Few developing countries can produce a surplus of food in good years, which might tide them over during bad years. As a result, these countries are extremely vulnerable to fluctuations in the climate. Unless constraints are placed on the burgeoning human population through family planning and birth control, additional misery and death will occur—especially among the poor nations. Because of the destruction of the topsoil, these people have already surpassed the carrying capacity of their land; they cannot possibly survive without outside aid.

8

The Loss of Life

THE human race was brought into existence during the greatest biological diversity in the history of the planet. We have been blessed with a highly diverse and interesting biosphere, filled with more species than have lived during any other period of geologic time. As human populations continue to expand and alter the natural environment, however, they are reducing biological diversity to its lowest level since the great extinction of the dinosaurs and many other species 65 million years ago.

It would be a great tragedy if, though our neglect of the environment and wanton destruction of life, we would return the planet to a condition of low diversity as a result of the extinction of vast numbers of species. The Earth would then require millions of years to repair the damage wrought by human folly. We could instead work toward preserving our planet for future generations of mankind and for the rest of the living world.

GLOBAL ENVIRONMENTS

The world is fortunate to possess such a large number of different environments, which in turn support a wide variety of species. Wetlands are among the richest ecosystems in the world (FIG. 8-1). They support many species of plants and animals, including valuable fisheries. About 65 percent of the shellfish harvested in the United States rely on these areas for spawning and nursery grounds. Wetlands also function as natural filters, removing sediments and some types of water pollution. Furthermore, they protect coasts against storms and the serious erosion problems that accompany them. However, as sea levels continue to rise as a result of higher global temperatures brought on by the

Fig. 8-1. Wetlands in Dodge County, Wisconsin. These areas are essential for waterfowl and other animals.

greenhouse effect, 80 percent of the U.S. coastal wetlands could be lost by the middle of the next century.

The world's wetlands are rapidly disappearing, mostly from destructive human activities. Wetlands are drained worldwide to provide additional farmland. Nearly 90 percent of recent wetland losses in the United States have been for agricultural purposes. Presently, woodland wetlands are disappearing at the alarming rate of 1200 acres a day (FIG. 8-2). The urgent need to feed hungry populations is the major reason why developing countries are draining their wetlands. Short-term food production has replaced the long-term economic and ecological benefits of preserving the wetlands. This action is responsible for the loss of local fisheries and breeding grounds for marine species and wildlife. In many cases, the destruction of the wetlands is irreversible.

Although tropical rain forests cover only about 6 percent of the land surface, they contain 66 percent or more of the Earth's species. The plants and animals of the rain forests are continuously being crowded out by humans. These animals are becoming extinct at a rate of 100 species a day as a result of the destruction of their habitats and the pollution of their environments with herbicides, insecticides, and industrial pollution.

Fig. 8-2. *The dry bed of the 2300-acre Swan Lake, Walworth County, South Dakota. Once a waterfowl refuge, it dried out as a result of a drought that began in 1974.*

Already half of the bird species of Polynesia have been eliminated through overhunting and the destruction of native forests. Throughout the world, the extinction of species is thousands of times greater than the natural background rate of extinction before the appearance of human beings.

The coral-reef environment (FIG. 8-3) supports a larger number of plant and animal species than any other marine habitat. The key to this prodigious productivity is the unique biology of corals, which plays a vital role in the structure, ecology, and nutrient cycles of the reef community. Coral-reef environments have among the highest rates of photosynthesis, nitrogen fixation, and limestone deposition of any environment. The most remarkable feature about coral colonies is their ability to form massive calcareous skeletons, some of which weigh several hundred tons.

Coral reefs are centers of high biological productivity and their fish are a major source of food in the tropical regions. Unfortunately, the spread of tourist resorts along coral coasts in many parts of the world is adversely affecting coral production in these areas. Such developments are almost always accompanied by increased sewage dumping, by overfishing, and by physical damage to the reef itself from construction, dredging, dumping, landfills, and the destruction of the reef to provide tourists with souvenirs and curios.

In many areas, such as Bermuda, the Virgin Islands, and Hawaii, development and sewage outflows have led to extensive overgrowth of thick algae mats that kill the reef. The algae suffocate the coral by supporting the growth of oxygen-consuming bacteria, particularly in the winter when the algal cover on shallow reefs is very high. This growth kills the living coral and it eventually destroys the reef. Furthermore, increasing ocean temperatures, brought on by a possible greenhouse effect, are bleaching many reefs,

Fig. 8-3. Cocos Island, Guam, showing coral reef and lagoon. Note the refraction of waves around the reef projection.

turning the corals ghostly white as a result of the loss of algae in their tissues. The algae nourish the corals and this loss greatly endangers the reefs.

The Arctic tundra regions of North America and Eurasia are among the most barren landscapes on Earth. Tundra covers about 15 percent of the total land surface. It winds around the "top" of the world, north of the tree line and south of the permanent ice sheets. In many of the world's mountainous regions, alpine tundra lies above the tree line and below the alpine glaciers. The vegetation in the two regions has much in common, however. Both regions consist mostly of stunted plants, often widely separated by bare soil or rock.

Tundra is one of the most fragile environments in the world; even small disturbances can wreak great damage (FIG. 8-4). Cross-country vehicle tracks from oil exploration activities over 40 years ago are still visible today. When the polar front sweeps across polluted regions of the Northern Hemisphere (mostly in Europe and northwest Asia) during the winter, it removes atmospheric pollution and transports it to the Arctic

Fig. 8-4. A gravel road near the Umiat, which shows severe subsidence in permafrost in the Anaktuvuk district, northern Alaska.

regions, where it contaminates the once pristine skies. This pollution produces a phenomenon known as *Arctic haze*, which is often as bad as the air pollution in many American suburbs.

A major concern over increasing amounts of atmospheric pollution is greenhouse warming caused by trapping escaping thermal energy from man-made greenhouse gases. If the arctic tundra thaws out entirely as a result of higher global temperatures, which are amplified in the higher latitudes, the increased methane released into the atmosphere by the decomposition of plant material could cause a runaway greenhouse effect, which could raise global temperatures to lethal levels.

Deserts compose about 30 percent of the land surface (TABLE 8-1). Most of the world's great deserts lie in the subtropics in a broad band that runs roughly between 15 and 40 degrees latitude, north and south of the equator. In the Northern Hemisphere, a series of deserts stretches from the west coast of North Africa, through the Arabian peninsula and Iran, and into India and China. In the Southern Hemisphere, a band of deserts runs across South Africa, central Australia, and west-central South America. Because of natural and human activities, land areas are increasingly becoming desertified, adding 15,000 square miles of new desert each year.

The world's deserts are not only the hottest and driest regions on earth, but they are also the most barren environments. Most of the world's desert wastelands receive only a minor amount of rain during certain times of the year; other regions have gone for years without a single drop of rain. About 10 percent of the desert lands consist of sandy dunes that march across the desert floor, driven by the winds, that bury everything in their paths (FIG. 8-5). The deserts are not totally barren of life, but only the hardiest plant and animal species can survive under these harsh conditions. Many desert species have developed some unique survival strategies that are found nowhere else on Earth.

TABLE 8-1. Major Deserts of the World.			
DESERT	LOCATION	TYPE	AREA SQUARE MILES × 1000
Sahara	North Africa	Tropical	3500
Australian	Western/interior	Tropical	1300
Arabian	Arabian Peninsula	Tropical	1000
Turkestan	S. Central U.S.S.R.	Continental	750
North America	S.W. U.S./N. Mexico	Continental	500
Patogonian	Argentina	Continental	260
Thar	India/Pakistan	Tropical	230
Kalahari	S.W. Africa	Littoral	220
Gobi	Mongolia/China	Continental	200
Takla Makan	Sinkiang, China	Continental	200
Iranian	Iran/Afganistan	Tropical	150
Atacama	Peru/Chile	Littoral	140

Fig. 8-5. Compound star dunes in Gran Desierto, Sonora, Mexico.

HABITAT DESTRUCTION

As long as human populations remained small and the environmental impacts of human activities were negligible, other species were virtually unaffected. However, after the invention of agriculture, the human character fundamentally changed. The Industrial Revolution sparked another major change, which had profound global consequences. Therefore, populations soared and the human presence was felt by the rest of the living world.

The human race is growing so explosively and destroying the environment so extensively that species are perishing in tragically large numbers. If the present rate of extinction continues, sometime during the next century the number of species lost as a result of human activities could equal or surpass the great extinctions of the geologic past. With our burgeoning human populations and high levels of destruction and pollution (FIG. 8-6),

PHOTO BY MARK YENICHET, COURTESY OF U.S. NAVY

Fig. 8-6. One of hundreds of waterfowl that was cleaned up after being covered with oil following an oil spill near Huntington Beach, California on February 7, 1990.

other species are forced aside, allowing the rise of more hardy species, some of which might be very destructive and harmful. In other words, by destroying beneficial predators, we allow ''pests'' to flourish, thereby upsetting the balance of nature.

We are destroying the world's forests as such an alarming rate that recovery is nearly impossible. Tropical rain forests, especially those in the Amazon Basin of South America, are decreasing at an annual rate of about 40 million acres (an area of about the size of Georgia). Thus far, the world's tropical forests have been reduced by 55 percent.

The once-rich forest along the Atlantic coast of Brazil has been cleared to less than 1 percent of its original cover. The remaining tropical rain forests of the world (FIG. 8-7) cover an area that is only about the size of the United States.

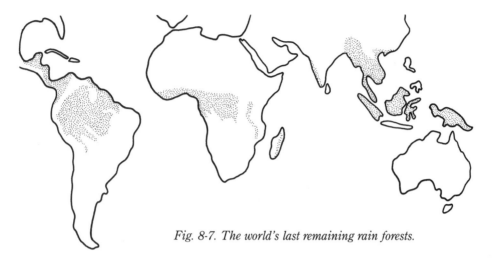

Fig. 8-7. The world's last remaining rain forests.

The rain forests are cleared mostly for agricultural purposes. About 15 percent are cut down for timber production, much of which is wasted through inefficient harvesting and milling methods. Some tropical forest fires are so massive and create so much smoke that orbiting spacecraft are unable to see the ground (FIG. 8-8). If global deforestation continues at the present rate, most of the world's rain forests will be gone before the middle of the next century.

Even the Swiss Alps are threatened with deforestation from air pollution that is generated largely by heavy tourist automobile traffic. Ironically, the tourist trade that the Alps have fostered is in the very process of destroying the mountains' great beauty.

Fifty percent of the forests of the world have already been cleared for agricultural purposes. Only about 15 percent of the once vast sea of forests that covered the United States still remains. What is left is rapidly being depleted for timber products. Each year, 60,000 acres of ancient American forests are being cut, mostly for lumber that is exported to countries on the Pacific rim.

Over 80 percent of Mexico's once vast tropical rain forest, the only one in North America, has been destroyed. Landless people conduct 75 percent of the global deforestation in a desperate search for food. Because of the scarcity of firewood, the only source of fuel for heating and cooking in impoverished regions, the forests of Africa are rapidly being depleted.

The process of deforestation is occurring over the entire world, but it is most insidious in the Amazon jungle of Brazil, where 20 million acres of forests (about the size of South Carolina) are burned each year. As much as 20 percent of the Amazon rain forest has already been destroyed. The cleared land mostly provides pasture for grazing cattle, whose beef products are exported to other nations. However, after only two or three years, the soil is eroded and robbed of its nutrients. Farmers are then forced to abandon their pastures and search for more land to clear.

NASA

Fig. 8-8. Heavy smoke from the clearing and burning of tropical forests, pastures, and croplands in the Amazon Basin of South America obscures the Earth from the Space Shuttle in 1989.

Modern methods of deforestation, through the widespread use of chainsaws, bulldozers, and giant timber harvesting machines, are contributing substantially to the rapid decline of the world's rain forests. Some rain forests, such as those in Hawaii, are harvested solely for electrical power generation—a tragic waste of a valuable resource. After the timber companies have taken the more desirable trees, unwanted trees and brush are burned and the bare soil, now denuded of all vegetation, is left unattended and is vulnerable to flash floods and erosion. When the rains come and the floods wash away the denuded soil, the area is completely decimated, providing no chance of recovery.

The harvest of forests, the extension of agriculture, and the destruction of wetlands destroys wildlife habitats, and speeds the decay of humus, discharging huge amounts of carbon dioxide into the atmosphere. Also, agricultural lands, which also produce carbon dioxide when they are cultivated, do not store nearly as much carbon as the forests that they replace. The clearing of land for agriculture, especially in the tropics, is the largest source of carbon released into the atmosphere from the biota and soils.

DISRUPTED FOOD WEBS

Upwelling ocean currents, near the coasts and the equator are important sources of bottom nutrients, such as nitrates, phosphates, and oxygen. Microscopic plant life, called *phytoplankton*, thrive in the surface waters of the ocean, where sunlight can penetrate for photosynthesis. These tiny organisms reside at the very bottom of the marine food web and are eaten by predators, which are themselves eaten by progressively larger predators.

As sea life flourishes, it depletes important nutrients from the water, limiting further growth of marine plant and animal populations. Fortunately, scattered around the world are numerous upwelling zones of cold, nutrient-rich water that can support prolific booms of phytoplankton and other marine life. These areas are also of vital economic importance to the commercial fishing industry. However, fluctuations in the atmospheric and oceanic circulation systems can cause periodic shifts in the coastal upwelling zones—with potentially disastrous consequences for the world's food supply.

Phytoplankton live in the phototropic zone of the oceans where photosynthesis occurs. The phytoplankton are responsible for 80 percent of all oxygen production in the world. They also provide the basic food stock, upon which all marine life ultimately depends for its survival. Any interference by man with the delicate balance within this thin membrane of water can have dire consequences for the ocean as well as for the entire world.

In the Antarctic Sea, whales, fishes, squid, and sea birds feed on small, shrimplike crustaceans called *krill*. With the decline of whale populations as a result of overfishing, other krill-eating animals have shown large population increases in recent years. Antarctic seals have had a rapid increase in population that has outpaced any simple recovery from past overhunting. Similarly, the population of penguins (FIG. 8-9) seems larger than

Fig. 8-9. Penguins and other species in the Antarctic Sea compete for krill that would otherwise be eaten by whales, which are threatened with extinction.

can be explained by simple recovery from the slaughter of the 19th century. Population increases have also been documented for some sea birds.

Apparently, the removal of most of the large baleen whales, which strain krill through their baleen plates, has affected competition. Therefore, the depletion of the whales has resulted in a surplus of krill. The increased krill has in turn attributed to the rise in populations of other krill-eating species. Roughly one million tons of krill are harvested each year by humans, but very little is actually eaten directly, and most is ground into animal feed for beef cattle and other domestic animals.

Among the world's fisheries, some major fish supplies have collapsed as a result of overfishing. In the tropics, a substantial fishing industry in the Gulf of Thailand has had a dramatic change in the relative abundance of various species. This region has had a roughly tenfold decline in the catch of "good" fish as compared with "trash" fish (less desirable species). The composition of the catch is also changing toward smaller fish species, and even the average size of fish within the same species is smaller. Thus, some good fish are now regarded as trash fish because only small individuals are caught.

Overfishing tends to drive populations below the levels where competition is important in regulating population densities. Therefore, under heavy exploitation, species that can produce offspring quickly and copiously are given a relative advantage. Thus, the fleshy, good fish are replaced by coarser trash fish. It is not certain to what extent these changes are caused by shifts in fish populations, by changes in patterns of commercial fishing, or by environmental effects. What is apparent, however, is that if present trends continue, the world's fishery will be composed of increasingly more trash fish.

GLOBAL EXTINCTION

Extinction events of the past were caused by natural phenomena, including changing climatic and environmental conditions. The present extinction rate, however, is on the order of 10,000 times greater than it was before the appearance of humans. Today's extinctions, which are occurring at a rate of four species per hour, are forced extinctions resulting from destructive human activities.

If the present spiral of human population growth and environmental degradation continues out of control, possibly before the middle of the next century, 50 percent or more of the species living on Earth today will become extinct. As a possible prelude to global extinction, frogs and other amphibians, animals that have been in existence for over 300 million years, are disappering all over the world. These creatures might be sounding an early warning that the planet is in grave danger.

The human species is the only animal ever that could cause the extinctions of large numbers of other species. All plants and animals that are not directly beneficial to man will be forced aside as growing human populations continue to squander the Earth's space and resources and contaminate the environment. It is still not fully understood how the complex interrelationships among species and between them and their environments work. However, it is becoming more apparent that the continued destruction of large numbers of species will leave this world entirely different biologically than the one we presently occupy.

The African elephant and black rhinoceros, along with other large mammals (FIG. 8-10), are in danger of becoming extinct as a result of a greedy ivory trade and human encroachment into their environment. Africa was once a sea of wild animals surrounding a few islands of humanity. Today, however, that situation has been completely reversed. The African birth rate, which is the highest in the world, is expected to triple the population to 1.6 billion by the year 2025. This birth rate has the potential of tripling the present rate of destruction, leaving little hope for African wildlife.

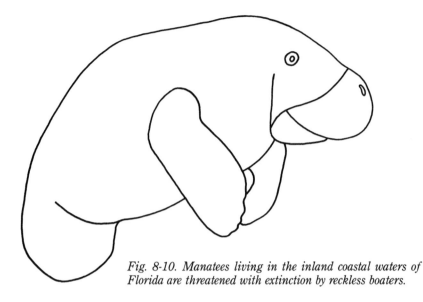

Fig. 8-10. Manatees living in the inland coastal waters of Florida are threatened with extinction by reckless boaters.

Certain exotic plants are rapidly becoming extinct. Some of these have important medicinal value. Over 50 percent of the pharmaceuticals are manufactured from natural herbs, most of which live only in the tropical rain forests. In the United States alone, as much as 10 percent of the nation's species of plants are destined for extinction by the end of this century. The native plants are at risk of extinction from habitat loss as a result of the destruction of the forests, the extension of agriculture, and the spread of urbanization. If current global trends continue, about 7 percent of all plant species on Earth will be extinct by the end of the century.

Tropical rain forests in the New World once covered an area of 3 million square miles. Today, that area has been reduced by about 30 percent. The tropical rain forests of Africa have been reduced by as much as 75 percent since 1960. Some nations, aware of the bleak future that awaits many species, are finally setting aside game preserves in an attempt to halt the tide of habitat destruction and extinction and perhaps safeguard the Earth for all of its inhabitants.

9

Hot as
a Greenhouse

MAJOR changes are occurring on Earth and it appears that our species is responsible for much of the climatic disturbances that beset our planet. The composition of the atmosphere has changed significantly; faster than at any other time in human history. In effect, human beings are conducting a global experiment by altering the environment. We are destroying the rain forests and pumping our pollutants into the air and water, thus changing the composition of the biosphere and changing the Earth's heat balance.

If we do not curb our insatiable appetite for fossil fuels and cease the destruction of the forests and their wildlife, by the middle of the next century, the world could be hotter than it has been in the past one million years. This heating would result from high levels of atmospheric carbon dioxide, along with other man-made greenhouse gases. Some of these artificial gases are extremely toxic and carcinogenic. Others are destroying the ozone layer, without which no life can exist on the Earth's surface.

THE GLOBAL GREENHOUSE

Record-breaking weather events have recently been occurring worldwide. The 1980s contained six of the seven hottest years of the last 140 years, even surpassing the Dust Bowl years of the 1930s (FIG. 9-1), which also featured record high temperatures. The decade of the 1990s promises to be warmer yet. These events might be symptoms of a global climate change caused by the chemical pollution of the atmosphere. However, the climatic variability is such that the strange weather could simply be a reflection of natural variation (TABLE 9-1).

Fig. 9-1. Buried machinery on a farm in Gregory County, South Dakota during the 1930s Dust Bowl.

TABLE 9-1. The Warmest, Wettest, and Windiest U.S. Cities.

EXTREME	LOCATION	AVERAGE ANNUAL VALUE
Warmest	Key West, FL	Mean temperature 78 deg. F.
Coldest	International Falls, MN	Mean temperature 36 deg. F.
Sunniest	Yuma, AZ	348 sunny days
Driest	Yuma, AZ	2.7 inches of rainfall
Wettest	Quillayute, WA	105 inches of rainfall
Rainiest	Quillayute, WA	212 rainy days
Cloudiest	Quillayute, WA	242 cloudy days
Snowiest	Blue Canyon, CA	243 inches of snowfall
Windiest	Blue Hill, MA	Mean wind speed 15 mph

Still no clear sign of climate change can be positively blamed on the greenhouse effect. Unknown moderating factors could cancel part of the greenhouse effect. Even pollutant aerosols might slow the greenhouse warming by reflecting sunlight back into space. Nature might "throw a monkey wrench" into climatic forecasts by erupting a large number of volcanoes in a relatively short period.

The mechanisms involved in greenhouse warming are not yet fully understood. However, the results of a steady rise in atmospheric carbon dioxide would probably be catastrophic without other moderating factors. These might include the absorption of excess carbon dioxide and heat by the oceans and green vegetation.

It still remains a mystery where the carbon dioxide produced by industrial activities is going. It appears that only about 40 percent of the carbon dioxide generated by the combustion of fossil fuels and by the destruction of the forests is accumulating in the atmosphere and absorbed by the ocean. Some of the excess carbon dioxide might be absorbed by terrestrial vegetation, where it would act like a fertilizer to stimulate growth. However, land plants cannot store as much carbon dioxide as the oceans can and this vegetation might reach capacity in the foreseeable future.

Within the space of 50 to 100 years, the world could become hotter than it was two million years ago (at the beginning of the present ice epoch) mainly as a result of high levels of atmospheric carbon dioxide (FIG. 9-2). The warming would be greatest at higher

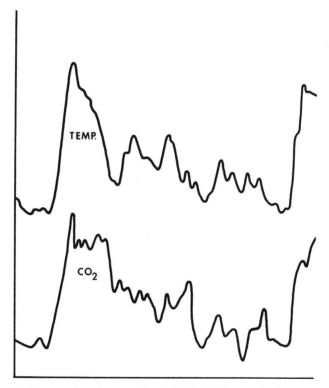

Fig. 9-2. The global temperature (top curve) and the level of atmospheric carbon dioxide (bottom curve) have kept in step over the last 160,000 years.

latitudes of the Northern Hemisphere, with the largest temperature increases occurring during the winter. Evaporation rates would increase, circulation patterns would change, and the weather would be much different.

The most unusual aspect of this warming trend, which amounts to an increase of over one degree Fahrenheit during this century, is its unprecedented speed. The present warming is 10 to 40 times faster than the average rate of warming following the last ice age. Between 14,000 and 10,000 years ago, the Earth warmed by perhaps 5 to 10 degrees. Although this rise in temperature is similar to the predicted increase for the greenhouse effect, the major difference is that it was spread over a period of several thousand years, rather than squeezed into a mere century.

If present warming trends continue, by the end of the next century, global temperatures could become as warm as they were 100 million years ago, during one of the hottest periods in geologic history when the dinosaurs reigned over the world. As a result of the higher temperatures, some areas, particularly in the Northern Hemisphere, would dry out and provide a large potential for massive forest fires. These possibilities have dire implications for us today. If greenhouse warming continues, major forest fires, such as those that devastated half of Yellowstone National Park in the summer of 1988 (FIG. 9-3), might become more frequent and consequently lose more forests and wildlife habitat.

Fig. 9-3. The forest fire that engulfed half of Yellowstone National Park, Wyoming in the summer and fall of 1988.

At the other extreme, the southern tropics could face severe flooding, which would erode cropland, displace people, and generally cause an ecological disaster of tremendous proportions. Rivers would be forced to carry more water than their channels could handle as they would race toward the sea, taking with them everything in their paths.

An increase in surface temperature, brought on by doubling the amount of atmospheric carbon dioxide, could have a global effect on precipitation. Subtropical regions would experience a marked decrease in precipitation, encouraging the spread of deserts. Increasing the area of desert and semidesert regions would significantly affect agriculture, which would be forced to move steadily poleward. Changes in precipitation patterns would have profound effects on the distribution of water resources needed for irrigation.

Along with diminishing rainfall, higher temperatures would augment evaporation; the flow of some rivers could decline by 50 percent or more, and others could dry out entirely. Major groundwater supplies would also be adversely affected, dropping water tables and drying out wells. Other areas would receive a marked increase in precipitation, causing extensive flooding that could be destructive to prime agricultural lands (FIG. 9-4).

PHOTO BY CAL OLSON, COURTESY OF USDA

Fig. 9-4. A flooded farmstead in the Red River Valley, North Dakota in 1975.

MAN-MADE GREENHOUSE GASES

Carbon dioxide emissions account for almost 60 percent of the annual human contribution of greenhouse gases. Carbon dioxide is responsible for 50 percent of the greenhouse warming, with water vapor and methane providing the bulk of the rest. The long-term increase in atmospheric carbon dioxide, as much as 25 percent since 1860, is the result of an accelerated release of carbon dioxide by the combustion of fossil fuels. Human activities, responsible for increasing the carbon dioxide content of the atmosphere (FIG. 9-5), promise to bring a general warming of the climate over the next several decades. Furthermore, the carbon dioxide lingers in the atmosphere for more than 100 years.

By the middle of the next century, the amount of atmospheric carbon dioxide could be twice the preindustrial level and global surface temperatures could rise as much as 5

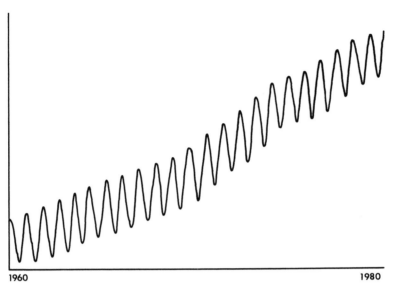

1960	1980

Fig. 9-5. Seasonal variation and the long-term increase in atmospheric carbon dioxide concentration.

to 10 degrees. This increase would dramatically change global precipitation patterns, enlarge the world's deserts, and drastically reduce agricultural output—conditions that could be disastrous on a planet that is rapidly becoming overcrowded.

Most of the carbon dioxide released into the atmosphere originates from the combustion of fossil fuels in factories, electricity generating plants, and motor vehicles, producing an average of about one ton of carbon for each of the world's 5.4 billion people annually. Americans release nearly six tons per person per year, amounting to 1.2 billion tons (about 25 percent of the world's total). By the year 2015, U.S. carbon dioxide emissions could increase by 50 percent.

Thirty percent of the carbon dioxide released into the atmosphere is from the destruction of tropical rain forests and the extension of agriculture, mostly by developing countries. The biota on the surface of the Earth and humus in the soil hold 40 times more carbon than the atmosphere. The harvest of the world's forests, the destruction of wetlands, and the extension of agriculture speed the decay of humus, thereby releasing vast quantities of carbon dioxide into the atmosphere.

Methane is the second most important man-made greenhouse gas. The atmosphere presently contains about one molecule of methane for every 200 molecules of carbon dioxide. However, methane is 20 to 30 times more effective per molecule at absorbing infrared radiation than carbon dioxide, which means that even small amounts released into the atmosphere can have a large effect. Methane production is outstripping that of carbon dioxide; it is increasing at about 1 percent per year, compared to about 0.5 percent for carbon dioxide. In the ensuing years, methane and other trace greenhouse gases might together contribute more to greenhouse warming than carbon dioxide alone.

Deforestation is rapidly increasing the number of termites. Presently, the Earth has about 3/4 ton of termites for every person. As deforestation escalates, that number could increase significantly. Termites ingest as much as 67 percent of all the carbon available on the land and about 1 percent of it is converted into methane.

In addition, cattle contribute substantial amounts of atmospheric methane during the digestive process. Anywhere from 5 to 9 percent of what a cow eats is converted to methane, representing about 5 percent of the total. Rice cultivation is another major contributor to the atmosphere load of methane. Also, natural gas, which is mostly methane, leaks from distribution systems and landfills.

THE CIRCULATION SYSTEM

Higher amounts of atmospheric carbon dioxide, along with other man-made greenhouse gases, principally methane, not only would increase surface temperatures, but they could also energize the atmosphere, invigorate the hydrologic cycle, and produce more intense storm systems. The general circulation of the atmosphere distributes heat from the tropics to the poles. If it was not for the circulation of air and ocean currents, the tropics would be much hotter and the higher latitudes would be much colder.

The Earth intercepts about one billionth of the Sun's energy, but only about 50 percent of the solar energy actually reaches the surface, where it mostly evaporates seawater. A great deal of thermal energy is required to vaporize water to form clouds.

When the clouds move to other parts of the world, they liberate thermal energy by precipitation, which distributes the ocean's heat around the globe. The type of clouds affect the global heat balance in other ways. Low stratocumulus clouds actually cool the Earth by reflecting sunlight off the cloud tops. High cirrus clouds warm the atmosphere by functioning as a greenhouse gas (FIG. 9-6).

Atmospheric circulation is also tied to the circulation of the ocean. The oceans play a vital role by storing the summer's heat and releasing it during the winter. The temperature difference from day to night and from summer to winter between the land and the sea is also responsible for the daily onshore and offshore winds and the seasonal monsoons that provide life-giving rains to half the world's people.

The transportation of heat by ocean currents plays a vital role in determining the climate. Ocean circulation is responsible for increased surface temperature at high latitudes, reduced snow and sea-ice cover, and lowered sensitivity to daily and yearly changes in atmospheric carbon dioxide concentrations. This circulation might explain why no strong link has been found between the rise in carbon dioxide levels and global temperatures.

Carbon dioxide enters the ocean from the atmosphere very slowly and at nearly a constant rate. Furthermore, the rate of absorption is only about 50 percent of the rate of carbon dioxide generation by the fossil fuel combustion. Normally, the atmosphere and the ocean are in equilibrium, with the amount of carbon dioxide being absorbed by seawater equal to that supplied to the atmosphere. As a result of increasing concentrations of carbon dioxide, however, humans are short-circuiting the *carbon cycle*, the flow of carbon through the biosphere.

The ocean's capacity for storing and moving vast quantities of heat (FIG. 9-7) surpasses that of the atmosphere by about 40 times. However, if climatic warming occurs too rapidly, the oceans might lose their equilibrium with the atmosphere. This loss could radically change circulation patterns and greatly affect the weather.

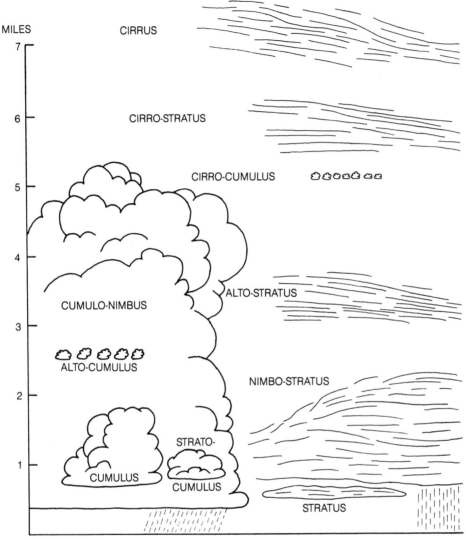

Fig. 9-6. Cloud types and altitudes. Stratocumulus clouds cool the Earth and cirrus clouds warm it.

ALTERED WEATHER PATTERNS

The weather results in a large measure from interactions between the ocean and the atmosphere. The troposphere, where most of the weather occurs, is a blanket of air that extends 6 to 10 miles from the Earth's surface and contains over 80 percent of the total mass of the atmosphere. The troposphere is thicker at the equator than at the poles, where the air is colder and therefore denser. The most striking feature about the troposphere is that it is in constant motion, moving air masses on a global scale. Warm air ascends at the equator, moves toward the poles, clashes with cold fronts, and produces storms.

Fig. 9-7. The ocean's heat transport system distributes warm water from the tropics to the higher latitudes. This system moderates the climates in both regions.

An increase in global temperatures, brought on by greenhouse warming, could energize the atmosphere to such an extent that storm systems would become much more violent. Furthermore, changing weather patterns as a result of instabilities in the atmosphere could make deserts out of once productive areas, while other regions would be drenched and have severe floods and soil erosion.

The potential larger-than-normal seasonal temperature variations and higher atmospheric moisture content would produce storms of unprecedented proportions. Dry winds of tornadic force would create gigantic dust storms and turn day into night. Tornadoes, hailstorms, thunderstorms, and lightning storms would increase in number and intensity. Numerous immense hurricanes would prowl the oceans and charge headlong into heavily populated coastal areas (FIG. 9-8), causing tremendous property damage and great loss of life.

Regions lying 30 degrees on either side of the equator can expect dramatic shifts in precipitation patterns as the world continues to heat up. The seasonal winds of the monsoons, which bring life-sustaining rainfall to half the people of the world, affect the continents of Asia, Africa, and Australia. If the monsoons would fail as a result of a climatic disturbance brought on by greenhouse warming, millions of people would be at risk of starvation. Major disruptions, such as the Kuwaiti and Iraqi oil-well fires, might also substantially affect Asia's monsoons. As the world's population continues to grow out of control, a major drought caused by the failure of the monsoonal winds could become the most horrendous human tragedy the world has ever known.

The central portions of the continents, which normally experience occasional droughts (FIG. 9-9), could become permanently dry wastelands. The soils in almost all of Europe, Asia, and North America would become drier, requiring upwards of 50 percent more irrigation. Expected rises in temperatures, increased evaporation, and changes in rainfall patterns would severely limit the export of excess food to developing nations during times of famine. Since the total heat budget of the Earth (FIG. 9-10) does not change significantly from year to year, areas that experience drought could have counterparts in areas that become unusually wet.

Fig. 9-8. Hurricane Diana off the North Carolina coast.

	0 – 10
	10 – 20
	20 – 30
	30 – 40
	OVER 40

Fig. 9-9. Rainfall amounts in the United States could drop in the interior portions of the country as a result of global warming.

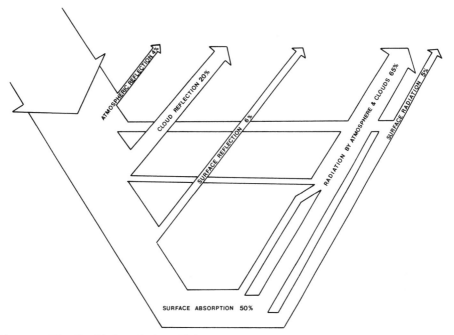

Fig. 9-10. The Earth's heat budget. Thirty percent of the solar radiation is reflected back into space before heating, 50 percent reaches the surface and is reradiated into space, and the rest is reradiated into space by the atmosphere.

THE RISING SEA

Coastal regions, where 50 percent of the world's people live, would feel the adverse effects of rising sea levels as the ice caps melt under increasing ocean temperatures. If the present melting continues, the sea could rise as much as one foot by the year 2030. For every foot that the sea level rises, up to one thousand feet of shoreline would disappear. This could cause large tracts of coastal land to be lost, along with shallow barrier islands and coral reefs. Low-lying fertile deltas that support millions of people would disappear. Delicate wetlands, where many species of marine life hatch their young, would vanish. Vulnerable coastal cities would have to move further inland or build protective walls against a raging sea.

The present rate of sea level rise is up to 10 times faster than 50 years ago. In most temperate and tropical regions, the sea level is rising at a rate of about 0.25 inch per year. Most of the increase appears to result from the melting of the ice caps. There also appears to be a greater number of icebergs, which calve off glaciers entering the sea. Furthermore, the icebergs seem to be getting larger. Alpine glaciers appear to be melting as well, mostly caused by global warming.

The increase in global temperatures also causes a thermal expansion of the ocean, increasing its total volume. Thermal expansion has raised sea levels as much as two inches over this century. A further rise in global sea levels would alter the shapes of the continents. The rising sea could also increase the risk of coastal flooding from high tides and storms. This problem is especially distressing for millions of people living on low-lying fertile deltas, such as the Bay of Bengal. This region was devastated by a gigantic cyclone on April 30, 1991, that was responsible for the deaths of 200,000 people.

The first possible signs that rising global temperatures have started to warm the ocean were revealed in satellite measurements of the polar sea ice, which shrank by as much as 6 percent during the 70s and 80s. Sea ice covers most of the Arctic Ocean and forms a frozen band around Antarctica (FIG. 9-11) during the winter season in both hemi-

Fig. 9-11. An iceberg and drift ice in the Antarctic Sea.

spheres. If global warming continues to melt the polar sea ice, the number of microscopic organisms would be reduced and the marine animals that feed on them would suffer as well.

THE BIOLOGICAL PRICE

In the past, climatic changes were slow enough for the biological world to adapt. However, today's climatic changes are far too rapid, possibly causing large numbers of species to become extinct. Global warming would be hardest on plants because they are directly affected by changes in temperature and rainfall. Forests, especially game preserves, would become isolated from their normal climate regimes, which would continue to move poleward as the planet gradually warms up.

The effects of global warming could last for centuries, during which time, forests would move to higher latitudes. Other wildlife habitats, including the Arctic tundra, would disappear entirely. One horrifying possibility is that the thawing of the Arctic tundra would release massive quantities of methane and create a runaway greenhouse effect.

Many species would be unable to keep pace with these rapid climatic changes. The warming would rearrange entire biological communities, and cause the extinction of many species. Those commonly called pests would replace other species and virtually overrun the landscape. High levels of carbon dioxide, which acts as a fertilizer, favor the growth of weeds. It would also be a heyday for parasites and pathogens, which could pave the way for an influx of tropical diseases in the temperate zone. The culminating effect would be a diminished species diversity worldwide, which would change the Earth into a totally alien planet.

10

The Satellite Solution

ADVANCES in remote sensing systems and computer technologies have provided mankind with new tools for monitoring and understanding the changes occurring on Earth. *Remote sensing* is widely used by biologists, geologists, geographers, agriculturalists, foresters, and engineers for evaluating natural and agricultural resources. Remote sensing has also been used for assessing crop inventory and yield, forest pest damage, fire damage, and range conditions; mapping and monitoring vegetation; measuring air and water quality; and providing vital information on other conditions that are important to the environment.

Data, taken from satellites in orbit around the Earth, is processed and enhanced by computers and correlated with instruments on the ground. Ozone depletion, carbon dioxide buildup, and acid rain have dramatized the need for international cooperation because pollution does not recognize national borders. In order to deal with the problems, scientists must have accurate and up-to-date information provided by the most sophisticated and reliable instruments ever devised. Although remote sensing offers the technology and a better perspective from which to study the Earth on a global scale, ultimately it is the scientist who must interpret the data and the individual who must act upon that interpretation.

POLLUTION FROM SPACE

As industrialization spreads throughout the developing countries of the world, so does pollution, an unfortunate byproduct of prosperity. The richest nations are also plagued with polluted air and water, contaminated soil and groundwater, and serious sewage, garbage, and toxic-waste disposal problems. Marine pollutants, such as raw sewage, industrial effluents, and oil slicks, can be detected by a variety of remote sensors on satellites.

Oil slicks are readily detectable by satellites because they have a higher index of refraction than the surrounding water does at ultraviolet and visible wavelengths. In the thermal infrared range, oil has different radiation characteristics than water. Oil slicks dampen small waves and thereby reduce the amount of radar backscatter, which can be detected on satellite radar imagery. With their continual surveillance, satellites can track the drift and dispersion of oil slicks for cleanup efforts.

A variety of remote sensors on Earth resource satellites (FIG. 10-1) can detect other surface pollutants, such as raw sewage and industrial effluents. The effects of dumping large quantities of wastes, such as sewage and industrial acids, into the ocean is observed by satellite and airborne sensors. Investigators use satellite imagery to differ-

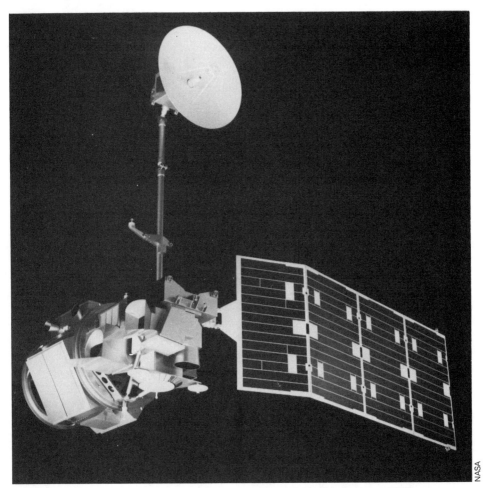

Fig. 10-1. An artist's rendition of a Landsat-D Earth resource satellite.

entiate between different types of wastes that are dumped into the ocean and to determine the drift and dispersion.

When used in conjunction with measurements taken on the water surface, satellite sensors can greatly improve the monitoring of certain pollutants. Stream sediment load and suspended particulates are also highly reflective. Consequently, sediment-laden rivers and estuaries appear lighter on satellite multispectral imagery. Not all pollutants are detected by satellite. They can be deduced by their association with other materials, however. Investigators can differentiate the types of wastes dumped into the ocean and determine their drift and dispersion rates. When used in conjunction with measurements taken on the surface, satellite sensors can enhance the ability to monitor certain pollutants in the water.

Air pollution around large cities and heavily industrialized areas is detectable on satellite imagery, either directly or indirectly, by its effects on the atmosphere. Organic compounds in the air interact with moisture and sunlight to produce photochemical smog, which is prevalent in many urban areas. Remote sensors can measure the concentration and movement of air pollution near large cities, the chemical emissions from industrial plants, and other chemicals in trace amounts in the atmosphere.

Atmospheric constituents, such as water vapor, carbon monoxide, and ozone, have been successfully measured from space by satellites such as NOAA's Nimbus 7 (FIG. 10-2). Satellites have also detected substantial reductions in the ozone layer over the polar regions, which are believed to be caused by chemical pollutants. The degree to which thermal radiation, emitted at the Earth's surface, is attenuated during its passage through the atmosphere is directly related to the concentrations of these gases. Furthermore, any increase in atmospheric carbon dioxide can increase the opacity of the atmosphere, thus altering the spectral distribution of outgoing energy.

Another remote-sensing technique for measuring atmospheric properties is by *lidar*, an acronym for *l*ight *d*etection and *r*anging. A short laser pulse is transmitted through the atmosphere and a portion of the radiation is reflected from atmospheric constituents, such as air molecules, clouds, dust, and aerosols. The interaction of the incident laser energy with these substances induces changes in the intensity and wavelength, depending on their concentration. As a result, information on the composition and physical state of the atmosphere can be deduced by analyzing the lidar data. In addition, the distances of the substances that interact with the laser beams can be determined by the delay of the backscattered radiation.

One of the most important uses of lidar is measuring the movement and concentration of air pollution near urban areas, the chemical emissions from industrial plants, and trace amounts of other chemicals in the atmosphere. It is also used to track the global circulation of volcanic ash emitted into the atmosphere by volcanic eruptions, such as the 1982 eruption of El Chichon in Mexico (FIG. 10-3). The climatic effects from this volcano, the dirtiest of this century, were felt over the entire world. Had it not been for remote sensors, including those carried on satellites, the effects of volcanoes on the climate might have been missed.

We might soon feel the adverse climatic effects of air pollution, especially carbon dioxide, particulate matter, acid vapors, and halocarbons. In order to prepare for such an uncertain climatic future, climatologists simulate the global climate with supercomputers. With mathematical equations to represent the interaction of the ocean, atmo-

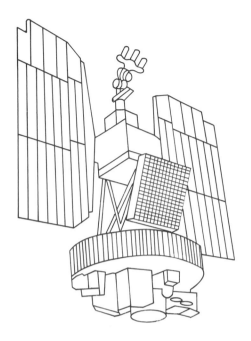

Fig. 10-2. The Nimbus-7 weather satellite.

NOAA

Fig. 10-3. The April 1982 eruption of El Chichon in Chiapas, Mexico from NOAA weather satellite.

sphere, and land, the computers can calculate how the climate will evolve in accordance with certain physical laws. The mathematical models cannot simulate all the complexities of the real climate, however; they can only reveal the consequences of certain actions.

OVERSEEING THE SEA

Dramatic changes in the climate can greatly affect our lives and the economy. Therefore, the world would benefit substantially from improved climate predictions. If the level of atmospheric carbon dioxide continues to increase, resulting from the combustion of fossil fuels and the destruction of forests and wetlands, global temperatures could be raised as a result of the greenhouse effect. These temperatures, in turn, could adversely affect the world's climate. They could also drastically affect the hydrologic cycle, which plays a critical role in sustaining life on Earth.

Improved climatic predictions depend on a better understanding of the effects that the ocean has on the climate. Space technology can provide the necessary tools for studying the ocean and its interaction with the atmosphere on a global basis (FIG. 10-4).

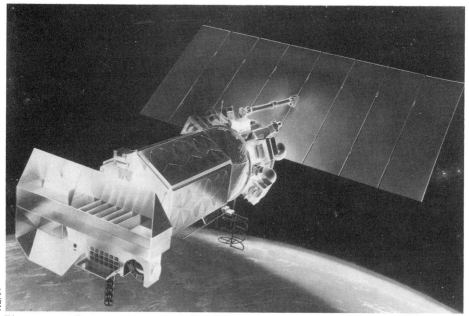

Fig. 10-4. A Tiros-N satellite that is designed for weather observations, ozone mapping, and Earth radiation budget measurements.

However, in order to understand this air-sea interaction, ocean currents, eddies, surface winds, and radiation must be measured simultaneously worldwide. Such measurements can only be obtained by sensors on satellites, complemented by measurements taken on the surface. This information is fed into powerful computers to help analyze and model the data in order to provide a timely assessment of the overall climate on the planet.

Radar satellites take altimeter measurements of the ocean topography (FIG. 10-5), which is affected by winds, currents, and gravity. The data can provide a better under-

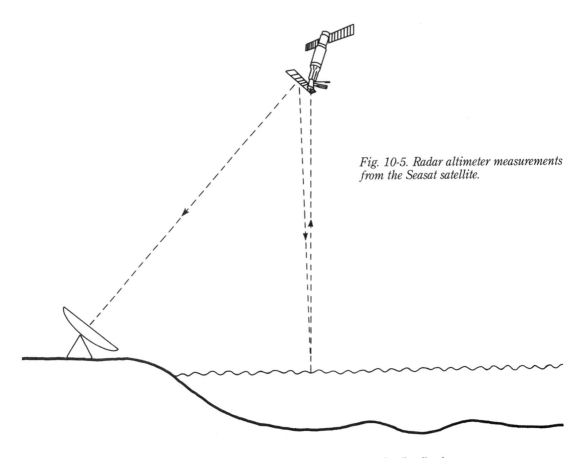

Fig. 10-5. Radar altimeter measurements from the Seasat satellite.

standing of the global oceanic circulation, which plays a pivotal role in distributing temperatures and nutrients around the world. Taking measurements of wind and ocean currents from satellites can give investigators a better understanding of the air-sea interactions that are central to climate predictions for the protection and control of the environment.

Satellite measurements of global biological activity, particularly the near-surface marine ecology, are essential for understanding the ecological cycles of the ocean. Satellite measurements of the ocean color can determine the chlorophyll content of the surface water layers, which is a direct indication of the primary sea production.

Phytoplankton (small, simple organisms in the surface waters of the ocean) are responsible for nearly all marine photosynthesis. They are the primary producers in the sea and they occupy a key position in the marine food chain, upon which all marine life ultimately depends. Because of the critical role that phytoplankton play in the marine ecology, which covers 70 percent of the Earth's surface, it is essential that accurate methods of assessing global marine primary production are developed. A complete understanding of phytoplankton distribution is critical for evaluating the general health of the seas and therefore, the health of the entire planet.

Remote sensing of the ocean color by satellite can provide comprehensive information on global marine primary production. The surface waters of the ocean change color in response to the amount of suspended matter, such as phytoplankton, sediment, and pollutants. Some parts of the ocean appear deep blue on satellite imagery because they lack phytoplankton or suspended sediment, and the color is caused by the optical qualities of seawater alone. By comparison, the waters of the North Atlantic appear green because they are rich in phytoplankton and are among the most productive areas in the world.

Swirling masses of water, called *eddies* or *gyres*, play an important role in mixing the waters of the ocean, just as tropical cyclones mix the atmosphere and provide a somewhat uniform temperature distribution throughout most of the Earth. Eddies are detected by satellites because of differences in their temperature, density, and chemical and biological composition, as compared to the surrounding water. These differences might manifest themselves as a change in surface roughness, as sea surface temperature differences, or as differences in color because of the phytoplankton content of the seawater.

One of the most important measurements of the ocean that can be made from space is surface wind. Backscatter from small wavelets and other surface features that are correlated with the wind are revealed on satellite radar imagery. Each point on the ocean is measured from two different aspects, which give the wind speed and direction. These measurements are constantly compared with and agree reasonably well to ship and buoy wind reports.

The surface winds are responsible, in most part, for generating the ocean's currents, which can be detected by radar altimetry. When a surface current is flowing, the Earth's rotation causes the ocean surface to tilt slightly, at right angles to the direction of flow. Major currents, such as the Gulf Stream, can produce a topographic relief on the ocean surface by as much as three feet. With further refinements, radar altimetry is capable of providing valuable information on global currents, which are important for distributing the ocean's heat and controlling a major part of the Earth's climate.

SURVEYING THE SURFACE

The growth rate and general health of the forests in many parts of the world have been declining over the past several decades. Many human-related factors are at work to destroy the forests. Detecting, identifying, and quantifying symptoms of forest decline with satellites and other orbiting vehicles, such as the Space Shuttle and the proposed Space Station (FIG. 10-6) are most urgent. Such observations provide investigators with the means to assess and to monitor forest destruction on a global scale, which would in turn provide governments with much needed information to help reduce forest destruction.

Deforestation assessments are made by comparing satellite images that are taken at different times. Vegetation stress generally reveals itself as changes in leaf structure, chlorophyll content, and water content, all of which produce spectral signatures (FIG. 10-7) that can be detected in computer-enhanced satellite multispectral imagery. The technique for measuring deforestation is basically quite simple. Healthy vegetation absorbs red light and reflects infrared light, but bare ground reflects red light and absorbs infrared light.

When a patch of forest is cut, the change in satellite imagery is clearly visible. By comparing images taken at different times, the change can be measured with considerable accuracy. In addition to mapping and monitoring forest decline and destruction,

Fig. 10-6. An artist's concept of the Space Shuttle orbiter undergoing servicing at the Space Station.

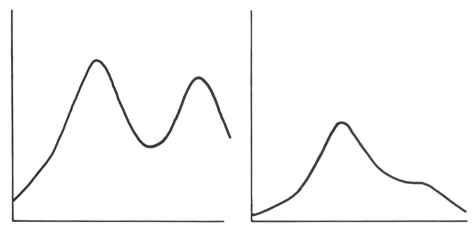

Fig. 10-7. Infrared signatures from a healthy tree (right curve) and one that is dying (left curve).

future satellite systems might allow investigators to detect spectral fingerprints that are associated with specific causes of deforestation, such as disease, infestation, drought, and expanding agriculture.

As forested and agricultural lands give way to continued urbanization, the change in spectral characteristics is readily discernible on multispectral satellite imagery. Up-to-date satellite imagery of land use is becoming increasingly necessary for planners to come to grips with the ever-growing problems of overpopulation on a planet with finite resources. The gathering of accurate data (which satellites can help provide) and intelligent planning will ensure the best possible use of these valuable resources.

In urban areas, satellite imagery can identify different levels of habitation, such as the central business district with a high density of buildings, dense residential areas with grass cover, and sparse residential areas with a moderate growth of trees. In many large American cities, a central core is surrounded by a prosperous and growing suburban region. By comparing imagery taken at various times, urban sprawl can be monitored as city limits continue to expand and consume the surrounding countryside.

FOCUSING ON FAMINE

Modern technology, with emphasis on satellite surveillance, can make the greatest contribution to world food security by analyzing the problem on a global scale. In many countries, the greatest threat to their security is not war, but ecological deterioration, which is accelerating at a furious pace. This manifests itself in vanishing forests and wetlands, top soil depletion, desertification, improper irrigation methods, overuse of groundwater, population pressures on limited food resources and other natural resources, and the effects that all these interrelated problems have on political and economic stability. It is senseless to apply stop-gap measures in order to keep up with the burgeoning human population, while destroying the very land that is needed to feed a starving world.

The success of world agriculture requires a systematic use of remote sensing data to monitor crop development. Too much is at stake to leave the future to ordinary crop forecasting methods, which have proven to be unreliable. Often, food-aid for Africa has been mobilized too late to deal with food emergencies with consequential losses of lives and dramatic increases in the expense of relief efforts.

The 1984 famine in Ethiopia and Sudan spurred international organizations to improve drought forecasting in order to give early warning of food emergencies. In recent years, these agencies have turned to remote-sensing techniques to cut response time and save lives. If predicting famine was a simple matter of detecting drought, satellite imagery could easily provide the necessary early warning. However, the causes of famine lie much deeper, requiring a knowledge of economic, political, and social factors, which vary from country to country.

The Famine Early Warning System (FEWS) was developed in 1986 by the U.S. Agency for International Development (USAID). This program takes physical as well as human factors into account. The aim is to provide timely information that can be used by agency planners and government officials for meeting food emergencies in African countries, whose dependence on relief will continue to grow as long as human populations continue to grow and destroy the land.

Satellite images of the Earth can provide the necessary tools for monitoring deforestation, droughts, crop fluctuations, and other natural and man-made processes on a global basis. Another important agricultural application of satellite imagery is mapping land-use potential and monitoring water availability. Satellite imagery can delineate large plots of land, which are cross-checked at selected sites in the field for accuracy (FIG. 10-8).

PHOTO BY TIM MCCABE, COURTESY OF USDA-SOIL CONSERVATION SERVICE

Fig. 10-8. A researcher takes soil moisture readings with a neutron probe, which is used to plan irrigation cycles to maximize efficient water use.

Accurate satellite image interpretation can detect agricultural growing conditions, such as drought, disease, infestation, or frost damage. Furthermore, each crop type has its own unique *spectral signature* (its response to different wavelengths of light), which can be used for broad-area crop identification (FIG. 10-9). Once crops are identified, crop production can be assessed. The crop production statistics are vitally important for developing countries because an unexpected shortfall accompanied by a delay in food imports could cause famine and the loss of a great many lives.

For developing nations, remote sensing has become critically important for economic growth. Too often, the economic strides made by poor countries are quickly eaten by an increasing number of mouths. Remote sensing from aircraft and satellites can provide valuable information on available natural resources, which must be properly understood and managed if developing countries are to become self-sufficient.

The technology has been utilized by several developing countries, which have focused their developmental strategies on specific applications, such as those directly related to energy sources. Most countries have realized that energy development is fundamental to the development of other natural resources; this development represents an asset that must be properly managed and sustained. Therefore, a means for rapid and comprehensive mapping of this important asset must be found. Remote sensing has become that means for nearly every nation on Earth.

NASA

Fig. 10-9. A landsat view of the San Joaquin Valley, California that shows computer-enhanced cultivated fields (bottom picture) for crop identification and yield.

11

⚜

Cleaning Up
Our Act

THROUGHOUT the world, environmental protection is often secondary to other concerns, such as the economy, lost jobs, bankrupt businesses, decreased productivity, and a host of other economic woes. A clean and healthy environment seems to be a luxury nations can ill afford. Money is required to clean up the environment, and pursuing this goal could damage the economy. The general attitude is that the environment will eventually be cleaned up sometime. Unfortunately, the burden would have to be passed on to future generations, who will ultimately pay for the selfish, short-term economic gains of this generation.

ENVIRONMENTAL CLEANUP

In 1990, the United States spent over $100 billion (about 2 percent of the gross national product) to clean up pollution. If trends continue, by the year 2000, that amount could double. However, the government's contribution is only a percentage of the total cost, which is mostly borne by industry and consumers. Furthermore, the rising costs do not necessarily reflect an increasing amount of pollution, but that environmental regulations are getting tougher.

Americans generate nearly 600 million tons of hazardous waste and wastewater each year, requiring up to $80 billion for disposal and treatment. About 1200 sites are on the Environmental Protection Agency's national priority list for cleanup (FIG. 11-1). That number is likely to double by the beginning of the next century.

Routine monitoring of industrial waste lagoons and landfills has revealed that chemicals are not being contained, which requires testing of nearby water wells. Once the pol-

131

Fig. 11-1. Location of major dump sites and aquifers that are identified for cleanup by the Environmental Protection Agency.

lution is found, determining its extent is difficult and expensive. Pools of chemicals generally advance at the rate of the subterranean flow. The pollution might also chemically react with the porous rocks of the groundwater aquifer. In the absence of air, the toxic substances might break down differently than they would on the surface.

In order to track the pollution: the direction and speed of the groundwater flow; the location, thickness, and composition of the aquifer; and its ability to transmit and store water must be determined. This procedure is accomplished by drilling a series of water wells across the suspected area; work that is both expensive and time-consuming.

Cleaning an aquifer is a nearly impossible task. In the simplest cases, when the contamination is fairly localized, the pollutants can be pumped out of water wells. If the chemicals have spread over a wide area, they can be blocked or encircled with impermeable clay. However, these methods only work when the contamination originates from a single source and it covers a limited area.

If the contamination is irreversible, the only recourse is to treat the water at the well head. Any contaminated water can be rendered drinkable, but it is very expensive to clean large quantities of water. Most states that bear the burden of monitoring and safeguarding the groundwater cannot afford the enormous investment without federal government support.

Cleaning landfills and waste lagoons is also costly. One method used to clean up soil that has been contaminated by underground storage tanks is *bioremediation*, which uses microbes to eat up toxic wastes. Ultimately, improved methods of dealing with waste problems and better knowledge of the underground environment will help solve future problems. Unfortunately, for much of the nation's groundwater supply, past mistakes might have already made recovery too late.

Oil spills are highly visible, dangerous, and they attract a lot of public attention. A number of environmental processes determine the outcome of an oil spill. The U.S. response strategy for major oil spills includes the use of computer models to predict the course of the spill for contingency planning. With their continual surveillance, satellites can aid in cleanup efforts by tracking the drift and dispersion of oil slicks.

Combating an oil slick requires containment with floating booms or an absorbent material (such as straw), and a labor intensive cleanup if the oil comes to shore. Detergents and other chemicals can be used to disperse the oil. Unfortunately, many of these chemicals are toxic and can cause additional ecological damage. Cleaning up oil-soaked beaches requires the use of chemical dispersants or steam (FIG. 11-2), which can kill

PHOTO BY JILL BAUERMEISTER. COURTESY OF U.S. FOREST SERVICE

Fig. 11-2. A Forest Service biologist checks an area that was cleaned by Exxon workers using high-pressure hot water, following the 1989 Alaskan oil spill. The treatment removes oil, but it also affects plant and animal life on the rocks.

organisms that would have otherwise survived if left alone, letting natural cleansing agents take effect.

Often, nature does a better job of cleaning up an oil spill than people do. Certain microbes can metabolize the oil, but the process takes a long time. Scientists have had limited success with fertilizing oil slicks with nitrates and phosphates to encourage the growth of oil-eating bacteria. Some microbes have been specially grown with voracious appetites for crude oil. They produce fatty acids when digesting the oil, so the petroleum products are more water-soluble. The remaining fatty acids also serve as food for plankton and other organisms. The drawbacks of the process are that the dissolving oil shifts the environmental damage to deeper water levels, where it could have a residence time of up to a decade. The introduced organisms could also upset the local marine ecology.

The spread of oil slicks is affected by ocean currents, local winds, and tides (FIG. 11-3). The success of cleanup operations depends on speedy rescue operations, the

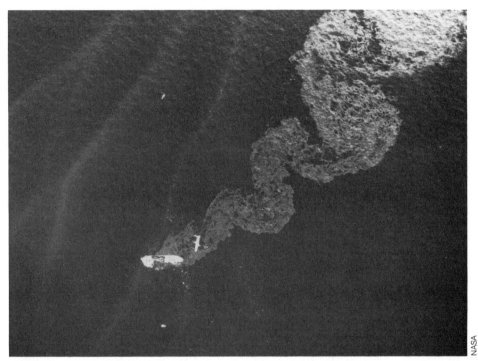

Fig. 11-3. Aerial view of the December 19, 1976, Argo Merchant *oil spill 28 miles off the coast of Nantucket, Massachusetts. Overflights of the oil-spill area aid in assessing the extent of the oil spill.*

availability of proper equipment, and accurate weather forecasts (particularly wind speed and direction). Data is collected on the dispersion, spread, and subsurface transport of oil slicks. The information is highly useful for determining the extent of the oil spill and the fate of the coastal ecology.

Efforts are being made to reduce acid precipitation from coal-fired plants by the installation of scrubbers on smokestacks to eliminate sulfur dioxide in addition to the combustion of low-sulfur coal. However, even if acid precipitation ended today, years or even decades would be required for the environment to recover. Moreover, many older plants built before 1975 and plants in other countries are not required to make these costly investments. Governments are slow to require mandatory emission controls, which are expensive and therefore might cause companies to lose their competitive edge.

Much of the coal from eastern underground mines (FIG. 11-4) has a high sulfur content. In order to meet clean air standards, many coal-fired plants are required to burn western coal, which is mined from massive open pits and transported over great distances. One method of transporting coal efficiently is by using a slurry pipeline, similar to the way petroleum products are transported. Many coal-fired generating plants could

Fig. 11-4. An underground coal-mining operation near Waltham, Massachusetts.

convert to natural gas, which is a much cleaner burning fuel, producing only 50 percent of the carbon as coal.

About 250 lakes and 25,000 miles of streams in the United States have pH values of 6 or lower. Many sensitive aquatic species can die from a long-term exposure to a pH of 6 or a short-term exposure to more acidic levels. One method of suppressing high acidic levels in lakes and streams is to seed them with *lime*, which is made from abundant limestone (FIG. 11-5) and can neutralize the acids. Furthermore, liming the watersheds that feed the steams and lakes is more effective than applying the lime directly.

The lime is usually in powder or pellet form and it is applied using aircraft. However, care must be taken so that potentially toxic over-buffering does not occur, which could be damaging as well. Halting the acid precipitation at the source, such as treating flue gases at coal-fired plants with lime, will clear up 50 percent of the U.S. acidic surface waters. The other 50 percent can be restored with the lime treatment.

Stopping ozone depletion might be a more difficult problem. Many companies, mostly in the United States, are turning away from refrigerants and solvents that contain chlorofluorocarbons (CFCs), which eat ozone molecules. However, older refrigeration units will still be in existence for some time and most other countries are slow to take similar action. The United States is one of the few nations that have completely banned the use of CFCs as propellants in spray cans.

Fig. 11-5. A limestone quarry in Franklin County, Alabama in 1932. Limestone composes about 10 percent of surface rocks.

Unfortunately, no suitable replacements for CFCs are available for the manufacture of foam plastics. In order to reduce consumption, these materials would have to be reduced. Another serious problem is that even if all ozone-destroying chemicals were banned immediately, the ozone layer would still be in danger of depletion because these substances linger in the upper stratosphere for long periods, possibly as much as a century or more.

COUNTERING GLOBAL WARMING

Human activities are mostly responsible for the long-term increase in the atmospheric carbon dioxide content. If present trends continue, a general climatic warming could occur over the next several decades. Although the mechanisms are complicated and not yet fully understood, the consequences of a steady rise in atmospheric carbon dioxide would probably be catastrophic if other factors do not moderate.

Unknown natural moderating factors could cancel part of the greenhouse effect. It is still unknown where all of the carbon dioxide produced by industrial activities is going. Only about 50 percent of the carbon dioxide that is generated by the combustion of fossil fuels and the destruction of the forests is accumulating in the atmosphere and in the ocean. However, the transfer of atmospheric carbon dioxide into the oceans is slow, and the seas can only process 50 percent of the excess carbon dioxide that is generated by humans.

The rest might be disappearing through the so-called *fertilization effect*. The extra carbon dioxide might allow plants to grow faster and larger, thereby removing more of the gas from the atmosphere. Unfortunately, high levels of carbon dioxide also encourage the growth of weeds. Planting more forests would soak up the excess carbon dioxide. The ancient forests (FIG. 11-6) would have to be saved as well because it takes about 200 years for the storage capacity of replanted forests to approach that of old-growth forests.

NATIONAL PARK SERVICE

Fig. 11-6. The General Sherman tree is a giant sequoia that is one of the largest trees on Earth and it is estimated to be between 2500 and 3000 years old. Sequoia National Park, California.

Marine single-celled plants, called *nannoplankton*, produce a gaseous sulfur that might help counter human-induced global warming by partially regulating the Earth's temperature. The sulfur gas emissions could increase the concentration of cloud-forming particles in the atmosphere, making clouds whiter, which would reflect more sunlight. This effect might lower global temperatures.

Other types of plankton might be encouraged to grow more vigorously by fertilizing the Antarctic Ocean with iron to help the organisms absorb more nutrients and enhance photosynthesis. They, in turn, would absorb more carbon dioxide from the atmosphere. The main source of iron in the ocean is dust blown from the world's deserts. One idea is to seed the ocean with iron, and a single supertanker load could trigger enough phyto-plankton growth to draw 2 billion tons of carbon out of the atmosphere. However, upwards of a century of iron fertilization might be required to reduce atmospheric carbon dioxide levels by 5 to 15 percent.

Through conservation methods, we can curtail some of the long-term effects of global warming. These measures would result in large part from the consequences of improved energy efficiency and the development of nonpolluting substitute energy sources (FIG. 11-7). Conservation would also make our world more environmentally healthy.

Fig. 11-7. The "Omega 24" system at the University of Rochester in New York is used to study laser fusion and related laser technologies.

One method of curtailing the ecological problems brought on by the greenhouse effect is through weather modification, such as using cloud-seeding methods (FIG. 11-8).

U.S. NAVY

Fig. 11-8. An aircraft seeding clouds in California.

Hurricanes, tornadoes, floods, and droughts are on-going threats to humanity. Man would like to modify these extremes of nature. With a fine-tuning of nature, deserts might one again bloom, swamps might dry up, monsoons might become more regular, and hurricanes might be steered away from populated coasts.

REFORESTATION

Growing urbanization sacrifices many trees to development. For example, Atlanta, Georgia, whose population has increased 30 percent every decade since 1970, has lost about 20 percent of its trees over the past 15 years. The loss of trees has also contributed to the urban heat island effect by replacing the cooling effects of trees with heat-retaining structures. By cooling the air, trees can also slow down heat-driven photochemical reactions that produce ozone from hydrocarbons and nitrogen oxides.

In order to slow the carbon dioxide buildup in the atmosphere while alternative non-polluting energy resources are developed, it is vitally necessary to plant additional trees. By doubling the volume of forest growth each year, the major fossil-fuel consuming nations could delay the onset of greenhouse warming for perhaps a decade or more. However, the destruction of the world's forests would have to be curtailed as well. A new forest of nearly 3 million square miles (roughly the size of the continental United States) is necessary to fully restore the Earth's carbon dioxide balance. This area would also be equal to all the tropical forests that have been cleared for agricultural purposes.

The additional trees would absorb the excess carbon dioxide discharged into the atmosphere by human activities, as well as produce oxygen (FIG. 11-9). Replanting per-

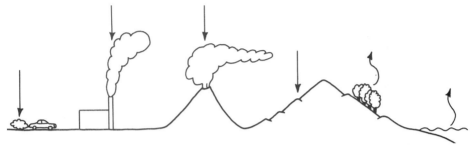

Fig. 11-9. Oxygen that is consumed by the combustion of fossil fuels and the oxidation of Earth materials is balanced by plant respiration on land and sea.

haps as many as 100 million trees would be required to remove about 18 million tons of carbon dioxide from the atmosphere each year. In the United States, lumber companies reseed the forests that they harvest to ensure a future supply of forest products. Unfortunately, replacing old forests can actually add carbon dioxide to the atmosphere, through the combustion, decomposition, and processing of wood products.

Even the fastest-growing loblolly pines still require 25 to 30 years to mature. However, in the past two decades, loblolly growth rates in the South have declined by as much as 15 percent as a result of air pollution (mostly from ozone, which is also responsible for about 90 percent of U.S. crop losses from air pollution). High ozone levels also reduce photosynthesis, increase cell damage, and remove carbon dioxide from giant sequoia seedlings, dramatically reducing their survival rate.

It is uncertain whether the human race has the two to three decades required for new trees to mature to solve the carbon dioxide problem. Furthermore, young trees do not absorb as much carbon dioxide as the trees that they replace. Also, for much of the world, deforestation has destroyed the topsoil. For many areas, replanting trees is no longer an option. However, numerous degraded lands throughout the world (FIG. 11-10) could be replanted in forests without conflicting with agriculture.

Many nations are setting aside forested areas in an attempt to halt the tide of deforestation and for use as game preserves. The amount of forested land in the United States and a few other countries has actually increased slightly in recent years. The United States Forest Service has taken millions of forested acres out of multiple use and set up wilderness areas. Unfortunately, the forests that surround these enclaves are still in danger of being destroyed.

IMPROVED FUEL EFFICIENCY

Natural gas is the most plentiful hydrocarbon energy source in the nation (other than coal). Switching to natural gas where possible would cut carbon dioxide emissions by 50

Fig. 11-10. Hummock and kettle topography in Glacier County, Montana.

percent. Electricity generating plants and even motor vehicles can take advantage of this fuel. Much of the natural gas throughout the world is simply vented into the atmosphere or burned off at well sites and refineries. Present reserves of natural gas can withstand steep increases well into the next century.

Coal is the most abundant hydrocarbon fuel, providing enough economically recoverable reserves to last another century. Unfortunately, it is also the dirtiest fuel in terms of emissions of carbon dioxide, oxides of nitrogen, and sulfur compounds, which can cause acid rain.

Acid rain is possibly the most well-studied pollution problem as a result of its severe damage to forests, streams, and lakes (FIG. 11-11). The source of the problem has been known for decades and the basic chemistry that turns industrial and motor vehicle exhausts into acid rain is well understood. However, governments are slow to require expensive mandatory emission controls. These controls might cause corporations to lose their competitive edge by having to make costly investments to clean up the environment.

Coal can be burned more efficiently by switching to pressurized fluidized bed boilers, which burn most of the pollutants, cut nitrogen oxide by 30 percent, and reduce sulfur emissions by nearly 95 percent over that of conventional power plants. Furthermore, pollution controls that are added to existing coal-fired power plants can cut nitrogen oxides and sulfur dioxides by nearly another 90 percent.

Instead of burning fuel in one plant to generate electricity and in another to manufacture products or to heat buildings, it is much more efficient to combine operations,

Fig. 11-11. A researcher from the Forest Service tests a lake for acidity levels.

rather than let the waste heat simply escape into the atmosphere. This procedure is called *cogeneration* and it can boost total efficiency by up to 90 percent and reduce air pollution by another 40 to 50 percent.

Industrial energy consumption per unit of production in the United States is much higher than that of other modern industrial nations. Improving efficiency could cut industrial air pollution by 50 percent. Improving insulation and using more efficient appliances and lighting in buildings and homes could reduce energy consumption and air pollution in those facilities by 50 percent or more (TABLE 11-1).

The transportation sector is responsible for producing 50 percent of the air contaminants from the combustion of fossil fuels. An improvement of five miles per gallon in auto mileage in the United States alone would cut carbon dioxide pollution by nearly 100 million tons a year. Energy efficient cars (FIG. 11-12) would reduce automotive carbon dioxide by up to 70 percent. Car pooling and mass transit would dramatically cut the smog in big cities. Developing more efficient systems for the transportation of goods, observing highway speed limits, and maintaining vehicles properly, will not only reduce pollution, but it will lessen our dangerous dependence on foreign sources of oil.

TABLE 11-1. Ten Everyday Ways to Clean Up the Planet.

Remember the three Rs: *Reduce*
Reuse
Recycle

1. Lower the thermostat 5 degrees Fahrenheit in winter and raise it 5 degrees in summer. Add extra insulation in walls and roofs and seal cracks around windows and doors. If every U.S. household did this, it could save the nation one million barrels of oil a day.

2. Turn the water heater temperature to 130 degrees Fahrenheit. This temperature is still hot enough to kill bacteria and it will reduce energy use by up to 50 percent. Adjust the refrigerator temperature controls to the proper operating range. Check door seals.

3. Preheat the oven in the shortest time possible. Place lids on pots to keep heat in and reduce cooking time. A pressure cooker consumes about 50 percent of the electricity that an oven uses. A microwave oven cooks faster using much less electricity. For cooking small amounts of food, use a microwave oven, toaster oven, or slow cooker.

4. Turn off all unnecessary lights. A 100-watt light bulb burning all day uses the energy of about 2 pounds of coal. One 100-watt bulb produces about the same light as two 60-watt bulbs and uses less energy. Use fluorescent lights. A 20-watt fluorescent bulb produces about the same light as a 100-watt incandescent bulb, and it lasts longer.

5. Don't leave the tap running. In one minute, one full gallon of water could be wasted. Turn off the tap while brushing teeth or shaving. If washing dishes by hand, wash all items in a sink of water and rinse under a slow-running tap.

6. Each flushing of a toilet requires about 5 gallons of water. Adjust the tank valve or put in a bottle full of water to save water. Installing a water-saving shower head could cut water use by 50 percent. Fix those leaks. One drip per second wastes 75 gallons per week.

7. Wash only full loads in dishwashers and clothes washers. Use the warm or cold cycle when washing clothes to save hot water. Use low-phosphate or phosphate-free cleaning products. Phosphates cause algae to grow, which kills other aquatic life. Chlorine in bleaches also ends up in streams and rivers and kills aquatic life.

8. Reuse paper grocery bags or use canvas bags. Buy beverages in returnable bottles and cans. Buy products in large containers, which save money and are thrown away less often. Pick products with containers made from natural materials, rather than plastic.

9. Use unbleached paper products. The bleaching process creates deadly dioxins. Use cloth items instead of paper for cleaning materials. Use cloth diapers rather than disposals.

10. Use separate containers for normal refuse and recyclables. Many garbage companies offer curbside pickup of recyclables, and some provide the service free. Collect newspapers, corrugated cardboard, glass, plastic, and aluminum cans for recycling. This measure could cut the amount of trash trucked to the nation's landfills by 50 percent.

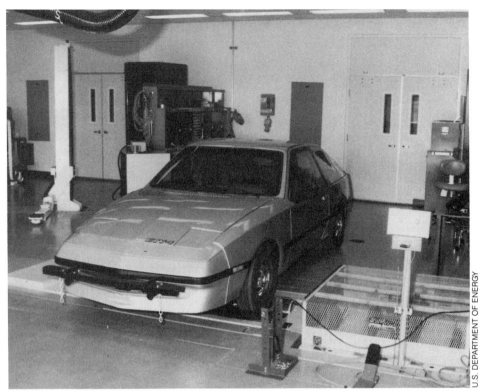

Fig. 11-12. An electric automobile is tested at the Idaho Laboratory Facility, Idaho Falls, Idaho.

12

Global Pollution

POLLUTION is so pervasive that it requires solutions on a global scale. Economic developments throughout the world are disrupting patterns of land and water use. Global destruction of forests and wildlife habitat, large-scale extraction and combustion of fossil fuels, and widespread use of man-made chemicals in industry and in agriculture appear to be altering cycles of essential nutrients in the biosphere and affecting the global climate. By pumping our pollutants into the air and water, we are unfavorably altering the composition of the global ecosystem. In effect, we are conducting a global experiment by rearranging the entire world.

THE CHANGING PLANET

We are irreversibly changing our planet by substantially altering the physical and the biological environment (FIG. 12-1). In 1987, the human population reached 5 billion and it is predicted that the Earth's carrying capacity could handle several times that many people. Unfortunately, in order to support such a large population, the rest of the living world would have to step aside to make room for additional agriculture, industry, urbanization, and other destructive activities.

The industrialization that made our modern civilization possible is in the process of polluting the very environment that we and the rest of the living world must inhabit. The release of cancer-causing chemicals and other hazardous substances into the atmosphere and ocean is far greater and more widespread than ever before suspected. Dangerous chemicals are seeping into the groundwater supply, contaminating the drinking

Fig. 12-1. Egypt's Nile Delta is one of the most intensely irrigated regions on Earth. What was once a desert has been turned into some 20,000 square miles of farmland that encompasses an area about the size of West Virginia.

water of many municipalities. Dumping toxic wastes into the ocean might cause irreversible changes in aquatic ecosystems. The disposal of ever-mounting stockpiles of nuclear wastes (FIG. 12-2) is highly crucial because they remain a hazard to life for thousands of years.

The tropical forests are being destroyed for agriculture and timber on an alarming scale. Every second, 1.5 acres of forest are destroyed. Along with the destruction of the forests, millions of plants and animals are killed. Through improper farming techniques, top soil is eroding at several times the replacement rate. Man-made deserts are spreading throughout the world and even once-fertile lands are becoming deserts. The fragile surface layer of deserts in the Middle East, which keeps the sands in place, are destroyed by the activities of war, promoting a major increase in dust storms and traveling sand dunes.

Huge amounts of carbon dioxide, produced by the combustion of fossil fuels and the destruction of forests, are released into the atmosphere. This carbon dioxide could raise global temperatures and adversely affect weather patterns, making areas either too dry or too wet for successful habitation. Combustion of fossil fuels also produces deadly acid rain, which is destroying forests, crops, fish, and much of the ancient beauty that was handed down from earlier civilizations.

Dramatic changes are caused by the improper use of land and water resources, the combustion of fossil fuels, and the use of chemicals in industry and agriculture. These

Fig. 12-2. On-site nuclear-waste storage at the Savannah River Plant, Aiken, South Carolina.

activities, if left unchecked, could permanently alter nature's delicate balance. The complicated interdependence that organisms have on each other and on their environment is not yet fully understood. What is becoming more apparent, however, is that if we continue to upset the balance of nature through our wanton negligence and waste, we will be left with a much different biological world than the one we originally inhabited.

CLIMATE CHANGE

If we fail to curb our insatiable appetite for fossil fuels and stop destroying the forests, by the middle of the next century, the world could be hotter than it has been in the past one million years as a result of high levels of atmospheric carbon dioxide and other man-made greenhouse gases. Some of these artificial gases are extremely toxic and carcinogenic. Others are destroying the ozone layer, which allows life to exist on the Earth's surface.

Average global temperatures have risen one degree Fahrenheit over this century (FIG. 12-3). If greenhouse gases continue to pollute the atmosphere, by the middle of the next century, global temperatures could rise as much as 5 to 10 degrees. The greatest amount of warming would occur in the higher latitudes of the Northern Hemisphere during winter. Most areas would experience summertime highs, well above the century mark, and every year new temperature records would be set. As a possible prelude to global warming, the decade of 1980s witnessed the six hottest years of this century.

The additional warming would cause atmospheric disturbances, which could produce more violent storms (FIG. 12-4) and higher death tolls—especially on a planet that is rapidly becoming overcrowded. Many areas in the Northern Hemisphere would dry out and become vulnerable to massive forest and rangeland fires. The fires would dump additional quantities of carbon dioxide into the atmosphere and create a vicious circle.

Changes in temperature and rainfall brought on by global warming would also change the composition of the forests. At the present rate of deforestation, most of the world's

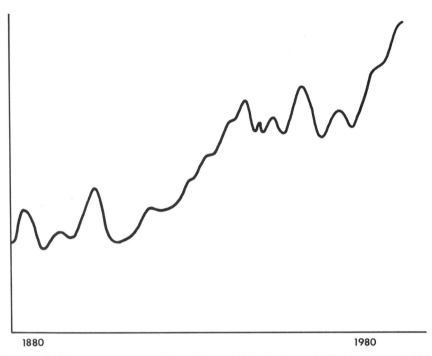

1880   1980

Fig. 12-3. Global temperatures rose from 1920 until 1940, dropped slightly between 1940 and 1970, and have been rapidly rising since.

U.S. ARMY CORPS OF ENGINEERS

Fig. 12-4. Tornado damage at La Place, Louisiana on December 6, 1983.

forests will be destroyed by the middle of the next century. The forests would be replaced by man-made deserts and sands would march across much of the landscape.

Evaporation rates would increase and change circulation patterns. Thus, rainfall would dramatically decrease in some areas and increase in others. In some regions, river flows would be reduced, and some rivers would dry out entirely. Other areas would experience sudden downpours, creating massive floods (FIG. 12-5). The continental inte-

Fig. 12-5. A bridge in the vicinity of Slidell, Louisiana that was washed out from a flood in the Pearl River on April 7, 1983.

riors that normally experience occasional droughts might become permanently dry wastelands. Vast areas of once-productive cropland could lose topsoil and become man-made deserts. As cropland diminishes, while populations continue to grow out of control, the greatest tragedy in human history could well be upon us.

Half of the human population lives in coastal regions (FIG. 12-6). These areas would feel the adverse effects of rising sea levels, as the ice caps melt under increasing ocean temperatures. If the present melting continues, the sea could rise as much as 6 feet by the end of the next century. Large tracts of coastal land would disappear along with wetlands, shallow barrier islands, and coral reefs.

The warming would rearrange entire biological communities and cause many species to go extinct. These extinctions would diminish species diversity, allowing weeds and pests to overrun much of the land. Since life itself controls the climate to some extent, it is uncertain what long-term effects a diminished biosphere would have on the world as a

Fig. 12-6. The development along the shorefront of Ocean City, Maryland. The distances between these buildings and the shoreline leaves little room for natural processes during storms.

whole. However, it is becoming more apparent that as our species continues to squander the Earth with a total disregard for its life-sustaining properties, the planet might be forced into another mode of operation that would not be favorable for human occupation.

ENVIRONMENTAL TERRORISM

Soon after the beginning of the Persian Gulf war in January 1991, the Iraqis deliberately dumped over one million barrels of crude oil from five supertankers at the Kuwaiti port of Mina. They also discharged another million or more barrels of oil from a tanker-loading terminal 10 miles offshore. The Allies were able to halt the flow of oil by bombing the inlet pipes that fed crude from several storage tanks to the undersea pipeline.

This act of environmental terrorism by the Iraqis created the world's largest oil spill, sending up to 3 million barrels of crude into the shallow and relatively enclosed waters of the Persian Gulf. The spill was about 12 times larger than the oil released off Alaska by the supertanker *Exxon Valdez*. Evidently, the Iraqis were using the oil slick to halt a U.S.

Marine amphibious assault on Kuwait. They might have also flooded the sea with crude oil in order to block the inlet pipes to a large Saudi desalination plant to the south, thus denying the Saudis a major source of fresh water.

This terrorist act could cause unprecedented damage to the ecology of the Persian Gulf for many years. Immediately, sea birds caught in the oil slick washed up on shore. Fish and shell fish also suffocated under the blanket of light, highly toxic crude oil. Tar balls that sink onto the seabed, kill bottom dwellers, destroy coral reefs, and ruin the great diversity of sealife that live on them. The spill could also destroy coastal wetlands and the multitudes of animals that depend on them. Sandy beaches might remain soaked with oil for several years.

In the wake of their retreat from Kuwait, the Iraqis set about 550 Kuwaiti oil wells on fire, blackening the skies with thick clouds of soot that turned day into dusk. In areas under the smoke clouds, the temperature dropped more than 20 degrees Fahrenheit lower than normal. By war's end, six million barrels of crude, about 10 percent of the world's daily consumption of oil, went up in smoke daily. Also, the fires sent 50,000 tons of sulfur dioxide and 100,000 tons of soot into the atmosphere every day. The balance of the one million tons of oil that went up in smoke each day was converted, for the most part, into carbon dioxide.

The smoke emissions, which were the worst ever recorded, caused severe and deadly pollution in Kuwait. Black rain loaded with oil soot and acids splattered the ground, blackening everything it touched. The local meteorological effects were comparable to those of nuclear winter, and the global effects will be felt for some time.

WAR AND POLLUTION

The most insidious pollution problem the world has yet to know would result from global nuclear and biochemical warfare (FIG. 12-7). The Earth's surface would become contaminated with radioactive fallout, toxic chemicals, and disease-ridden pathogens. Runoff would carry these poisons into the ocean, turning it into a vast graveyard. On land, the species would tragically die out as the Earth would diminish its ability to sustain life.

The world's nuclear arsenals contain roughly 20,000 megatons divided among some 50,000 warheads. This amount is equivalent to four tons of TNT for every man, woman, and child on Earth. Half of the human population, most of which live in the Northern Hemisphere, would die within the first few weeks following a full-scale nuclear war. The war would produce highly radioactive fallout that would continue to kill months and years later (TABLE 12-1). The fallout would be dispersed throughout the world—even those living in the Southern Hemisphere would not be spared from its deadly effects.

People living in major cities or near military targets would be inflicted with radiation burns as well as thermal burns from the fireball and from fires of combustible materials. Burning cities would fill with dangerous toxic fumes that would clog the air and choke survivors (FIG. 12-8). Forests near targeted areas would also catch fire from the thermal pulse generated by the fireball and burn out of control.

Past assessments of nuclear war only considered the damage to industry along with human casualties, but failed to account for the ecological damage to the Earth. Disease from decomposed, unburied corpses, both human and animal, would spread like wildfire. Some plants, including certain crops, have a particularly low tolerance to nuclear radia-

Fig. 12-7. An atomic bomb detonation at the Nevada Test Site on January 17, 1962.

tion and many animals have a much lower tolerance than humans. Pests, such as insects and rats, have a considerably higher tolerance to nuclear radiation. Without their natural predators, they would multiply rampantly, spreading diseases in epidemic proportions.

Soil and water would be heavily contaminated with radioactive isotopes and toxins. Sulfur and nitrogen compounds released into the air from burning cities would produce acid rain, which is highly destructive to plants and aquatic life. A large number of nuclear detonations would also destroy the ozone layer and expose any surviving inhabitants on the surface to dangerous levels of ultraviolet radiation from the Sun.

TABLE 12-1. The Effects of Radiation.

DOSE, REM	IMMEDIATE EFFECTS	LONG TERM EFFECTS	COMMENTS
0 - 50	No apparent injury Chromosome change	Genetic damage	Can cause birth defects and later cancers
50 - 100	Minor vomiting, depressed blood count, malaise, slight fatigue	Same as above	Symptoms begin in a few hours and may last a day or more
100 - 200	Vomiting more common, loss of appetite, sore throat, thirst, fever, minor diarrhea	Symptoms may reappear in two or three weeks	Less than 5% will die from radiation sickness
200 - 300	Above symptoms more pronounced, greater fatigue, sterility in both sexes	Gum, intestinal and skin bleeding, loss of hair, depletion of white blood cells cause infection	20% of the population will die within two weeks
300 - 400	Symptoms more severe and occur more rapidly, most victims require clinical treatment	Gastrointestinal disorders, kidney bleeding, loss of hair, leukemia and thyroid cancer	Death from dehydration and starvation, up to seven week hospital stay
400 - 500	Same as above only worse, all victims require clinical treatment	Skin burn and skin cancer, extensive internal bleeding, gastric ulcers, liver and lung disease	50% fatality, death in one month, ten week hospital stay
500 - 1000	Generally fatal, severe illness, vomiting will continue for several days, intense cramps and bloody diarrhea	Permanent sterility and loss of hair, cataracts, intense ulceration, nerve disorders, pneumonia	100% fatality without intense clinical treatment, extreme agony, death in two weeks
over 1000	Victim will go into spasms and convulsions, death in one to two days	None	Virtually all die, onset of coma within a few hours

Smoke emission from a full-scale nuclear war would be 300 million tons. Spreading the smoke evenly around the world would reduce the amount of sunlight reaching the ground by as much as 95 percent (FIG. 12-9). Heavier smoke would cover the targeted areas, where daytime would be as dark a a moonlit night. The detonation of large numbers of nuclear weapons at ground level would also inject several hundred million tons of soil and dust into the atmosphere.

The reduced sunlight from smoke and dust could cool the Earth by several tens of degrees Fahrenheit. This cooling could persist from several months to over a year. Thus, temperate zones would freeze in the middle of the summer. As a result of the cold and the dark, plants, which depend on warmth and sunlight, would wither and die from the onset of a sudden freeze. Therefore, any attempts at growing crops to survive would be futile.

Fig. 12-8. *A firestorm created by a nuclear detonation on a city.*

Fig. 12-9. *The July smoke line, above which sunlight is twilight or darker during the daytime three weeks following a global nuclear war.*

The large temperature difference between the land and sea would produce violent coastal storms. The runoff would carry huge quantities of radioactive poisons, toxic chemicals from destroyed cities, and disease-ridden organic matter into the ocean. In effect, the ocean would become a vast nuclear and toxic chemical dump. The continual darkness and poisons would kill primary producers in the ocean and cause tragic consequences up the food chain. The ocean would thus experience a cataclysmic die-out of species—far worse than the greatest extinction in the history of the Earth.

A DYING PLANET

Life constitutes a geological force that is missing on other planets. The evidence of biospheric processes in the Earth's history belongs to the field of *biogeology*, which covers the effects organisms have on geological activities. Life processes have been operating for at least 80 percent of geologic history. With this much time involved, it is not surprising that life caused some dramatic, far-reaching changes on the Earth.

It is also becoming more apparent that humans are upsetting the delicate balance of nature; if it should tilt ever so slightly, cataclysmic changes might result. Over a period of more than four billion years, the Earth has perfected a means of survival for all living things. Major readjustments have been followed by major extinctions, but life generally emerged from these tragic events brutally scathed, but not totally destroyed.

Modern-day extinctions, if compared to those of the past, would not span a period of millions or even thousands of years, but instead take place in merely a century. Large mammals, such as whales, manatees, elephants, and rhinos, appear to be going the same route as their large reptilian counterparts, the dinosaurs. This time the extinction pressure is not coming from natural causes, but instead by the hand of man.

Despite serious efforts to stop the extinction of certain species, the number of animals that are disappearing is far exceeding the natural background rate. If the present spiral of human population and environmental destruction continues out of control, sometime during the next century, 50 percent or more of the species living on earth today would be gone.

The extinctions of the past were caused by natural phenomena, as a result of major changes in the environment. Present-day extinctions are forced, caused by man's destructive activities. If the present rate of extinctions continues to rise as man's population and environmental destruction continues to climb, the Earth is placed in grave danger of losing its great species diversity. We should not delay in finding solutions to these immense problems. If we destroy the Earth (FIG. 12-10), we also destroy ourselves.

Fig. 12-10. The edge of a smoke cloud from massive forest fires in the Amazon Basin that obscured almost 30 percent of South America.

NASA

Glossary

absorption—The process by which radiant energy that is incident on any substance is retained and converted into heat or another form of energy.

abyss—The deep ocean, generally over one mile deep.

accretion—The accumulation of celestial dust by gravitational attraction into a planetesimal, asteroid, moon, or planet.

acid precipitation—Any type of precipitation with abnormally high levels of sulfuric and nitric acids.

adiabatic—Changes in temperature that occur within air masses as a result of pressure changes, which cause the masses to expand or contract without gain or loss of heat.

adsorption—The adhesion of a thin film of liquid or gas to the surface of a solid substance.

advection—The horizontal movement of air, moisture, or heat.

aerosol—A mass made of solid or liquid particles that are dispersed in air.

air mass—An extensive body of air whose horizontal distribution of temperature and moisture is nearly uniform.

airstream—A substantial body of air with the same characteristics flowing with general circulation.

albedo—The amount of sunlight that is reflected from an object.

alluvium—Stream-deposited sediment.

alpine glacier—A mountain glacier or a glacier in a mountain valley.

anticyclone—The circulation of air around a central area of high pressure that is usually associated with settled weather; pressure rises steadily when an anticyclone is developing and falls when it is declining.

aquifer—A subterranean bed of sediments through which groundwater flows.

ash, volcanic—Fine pyroclastic material that is injected into the atmosphere by an erupting volcano.

atmospheric pressure—The weight per unit area of the total mass of air above a given point. Also called barometric pressure.

benthic front—Ocean currents that flow along the bottom of the deep ocean.

biogenic—Sediments that are composed of the remains of plant and animal life, such as shells.

biosphere—The living portion of the Earth that interacts with all other geological and biological processes.

blocking high—Any high pressure center that remains stationary, effectively blocking the usual eastward progression of weather systems for several days in the middle latitudes.

calving—The formation of icebergs as a result of glaciers breaking off in the ocean.

carbon cycle—The flow of carbon into the atmosphere and ocean. The carbon is converted to carbonate rock, then it returns by volcanoes.

carbonate—A mineral that contains calcium carbonate, such as limestone and dolostone.

circulation—The flow pattern of a moving fluid.

coalescence—The merging of two or more colliding water droplets into a single, larger drop.

cold front—The interface or transition zone between advancing cold air and retreating warm air.

coastal storm—A cyclonic, low-pressure system that moves along a coastal plain or just offshore. When it causes north to northeast winds over the land and along the Atlantic seaboard, it is called a *northeaster*.

condensation—The process whereby a substance changes from the vapor phase to liquid or solid phase; the opposite of evaporation.

conduction—The transmission of energy through a medium.

convection—A circular, vertical flow of a fluid medium as a result of heating from below. As materials are heated, they become less dense and rise, while cooler, heavier materials sink.

convergence—A distribution of wind movement that results in a net inflow of air into an area, such as a low-pressure area.

coral—Any of a large group of shallow-water, bottom-dwelling marine invertebrates which build reef colonies in warm waters.

Coriolis effect—The apparent force that deflects wind and ocean currents, causing them to curve in relation to the rotating earth.

cyclone—The circulation of air around a central area of low pressure that is usually associated with unsettled weather. In tropical latitudes, it can refer to an intense storm that does not attain full hurricane status.

density—The amount of any quantity per unit volume.

dew—Water droplets that are caused by condensation of water vapor from the air as a result of radiation cooling.

dew point—The temperature to which air, at a constant pressure and moisture content, must be cooled for saturation to occur.

differentiation—The separation of solids or liquids according to their weight; heavy masses sink and light materials rise toward the surface.

diffusion—The exchange of a fluid substance and its properties between different regions in the fluid as a result of small, almost random motions of the fluid.

downwelling—A cold fluid sinks because it is heavier than the surrounding medium.

drought—A period of abnormally dry weather that is sufficiently prolonged for the lack of water to cause serious deleterious effects on agricultural and other biological activities.

dune—A ridge of wind-blown sediments usually in motion.

eolian—A deposit of wind-blown sediment.

estuary—Tidal inlet along a coast.

evaporation—The transformation of a liquid into a gas.

evolution—The tendency of physical and biological factors to change with time.

exosphere—The outermost portion of the atmosphere that is in contact with space.

fossil—Any remains, impressions, or traces in rock of a plant or animal of a previous geologic age.

frost—Ice crystals formed on grass and other objects by the sublimation of water vapor from the air at below freezing temperatures.

geothermal—Relating to the movement of hot water through the crust.

glacier—A thick mass of moving ice that occurs where winter snowfall exceeds summer melting.

greenhouse effect—The trapping of heat in the atmosphere principally by carbon dioxide.

groundwater—The water derived from the atmosphere that percolates and circulates below the surface of the Earth.

hazardous waste—Any pollutant that is particularly harmful to life, including toxic substances and nuclear wastes.

heat flow—Heat energy flows from hot towards cold at a rate or flux that is equal to the temperature gradient times the conductivity of the material in between.

high—An area of high atmospheric pressure within a closed circulation system; an anticyclone.

hydrocarbon—A molecule that consists of carbon chains with attached hydrogen atoms.

hydrologic cycle—The flow of water from the sea to the land and back, which plays a major role in cleansing the Earth.

hydrosphere—The water layer at the surface of the Earth.

ice age—A period of time when large areas of the Earth were covered by glaciers.

iceberg—A portion of a glacier that is broken off upon entering the sea.

ice cap—A polar cover of ice and snow.

infrared—Heat radiation that has a wavelength between red light and radio waves.

insolation—All solar radiation that impinges on a planet.

interglacial—A warming period between glacial periods.

Intertropical Convergence Zone—The axis along which the northeast trade winds of the Northern Hemisphere meet the southeast trade winds of the Southern Hemisphere.

inversion—A departure from the usual decrease or increase, with altitude, of the value of an atmospheric property.

jet stream—Relatively strong winds concentrated within a narrow belt that is usually found in the tropopause.

landslide—The rapid downhill movement of earth materials that is often triggered by earthquakes.

lapse rate—The decrease of an atmospheric variable (usually temperature) with height.

latent heat—Heat absorbed when a solid changes to a liquid or a liquid to a gas with no change in temperature, or heat released in the reverse transformation.

limestone—A sedimentary rock that mostly consists of calcite.

low—An area of low atmospheric pressure—a cyclone or a depression.

lysocline—The ocean depth below which the rate of dissolution just exceeds the rate of deposition of the dead shells of calcareous organisms.

mean temperature—The average of any series of temperatures observed over a period of time.

mesosphere—A region of the Earth's atmosphere between the stratosphere and thermosphere that extends 24 to 48 miles above the Earth's surface.

methane—A hydrocarbon gas that is liberated by the decomposition of organic matter.

monsoon—A seasonal wind that accompanies temperature changes over land and water from one season of the year to another.

nitrogen cycle—The flow of nitrogen from the atmosphere to living organisms, then back to the atmosphere when the organisms die.

oil spill—The dumping of crude oil on bodies of water, which is harmful to marine life and habitats.

outgassing—The loss of gas within a planet, as opposed to degassing, the loss of gas from meteorites.

ozone—A molecule that consists of three atoms of oxygen, which exists in the upper atmosphere and filters out ultraviolet light from the Sun.

permafrost—Permanently frozen ground.

permeability—The ability to transfer fluid through cracks, pores, and interconnected spaces within a rock.

phenology—The study of the times of recurring natural phenomena in relation to climatic conditions.

photosynthesis—The process by which plants create carbohydrates from carbon dioxide, water, and sunlight.

polar air—An air mass that is conditioned over the tundra or snow-covered terrain of high latitudes.

polar front—A semipermanent discontinuity that separates cold polar easterly winds and relatively warm westerly winds of the middle latitudes.

pollutant—Any substance that pollutes air or water, whether man-made or natural.

porosity—The percentage of a rock that consists of pore spaces between crystals and grains, usually filled with water.

precipitation—Products of condensation that fall from clouds as rain, snow, hail, or drizzle.

radiation—The process by which energy from the Sun is propagated through a vacuum of space as electromagnetic waves. A method, along with conduction and convection, of transporting heat.

reef—The biological community that lives at the edge of an island or continent. The shells form a limestone deposit that is readily preserved in the geologic record.

regression—A fall in sea level that exposes continental shelves to erosion.

relative humidity—The ratio of the amount of moisture in the air to the amount that the air would hold at the same temperature and pressure if it was saturated. Relative humidity is usually expressed as a percentage.

saltation—The movement of sand grains by wind or water.

saturated air—Air that contains the maximum amount of water vapor it can hold at a given pressure and temperature. Saturated air has a relative humidity of 100 percent.

soluble—A substance that dissolves in water.

storm surge—An abnormal rise of the water level along a shore as a result of wind flow in a storm.

stratosphere—The upper atmosphere above the troposphere, which is about 10 miles above sea level.

sublimation—A process by which a gas is changed into a solid or a solid is changed to a gas without passing through the liquid state.

subsidence—The descending motion aloft of a body of air, usually within an anticyclone. Subsidence spreads out and warms the lower layers of the atmosphere.

supercooling—The cooling of a liquid below its freezing point without it becoming a solid.

surfactant—A surface-active substance on top of a body of water that controls the diffusion of gases and other substances.

surge glacier—A continental glacier that advances toward the sea at a high rate.

temperature inversion—A layer of the atmosphere in which the temperature increases with altitude, as opposed to the normal tendency for temperature to decrease with altitude.

thermosphere—The outermost layer of the atmosphere in which the temperature increases regularly with height.

tide—A bulge in the ocean that is produced by the Moon's gravitational forces on the Earth's oceans. The rotation of the Earth beneath this bulge causes the sea level to rise and lower.

transgression—A rise in sea level that floods the shallow edges of continental margins.

tropical cyclone—A low-pressure area that originates in the tropics, has a warm central core, and often develops an eye.

troposphere—The lowest 9 to 12 miles of the Earth's atmosphere, which is characterized by a temperature that decreases with height.

tundra—Permanently frozen ground at high latitudes and high altitudes.

typhoon—Severe tropical storms in the western Pacific that are similar in structure to a hurricane.

ultraviolet—The invisible light that has a wavelength shorter than visible light and longer than X-rays.

virga—Wisps or streaks of water or ice particles, which fall from clouds and evaporate before reaching the ground.

viscosity—The resistance of a liquid to flow.

warm front—The boundary of an advancing current of relatively warm air, which is displacing a retreating colder air mass.

water vapor—Atmospheric moisture that is in the invisible gaseous phase.

Bibliography

THE POLLUTED PLANET

Carns-Smith, A. G. "The First Organisms." *Scientific American* Vol. 252 (June 1985): 90–100.

Holland, H. D., B. Lazar, and M. McCaffrey. "Evolution of the Atmosphere and Ocean." *Nature* Vol. 320 (March 6, 1986): 27–33.

Kasting, James F., Owen B. Toon, and James B. Pollack. "How Climates Evolved on the Terrestrial Planets." *Scientific American* Vol. 258 (February 1988): 90–97.

Lewin, Roger. "A Revolution of Ideas on Agricultural Origins." *Science* Vol. 240 (May 20, 1988): 984–986.

Lunine, J. I. "Origin and Evolution of Outer Solar System Atmospheres." *Science* Vol. 245 (July 14, 1989): 141–146.

Pilbeam, David. "The Descent of Hominoids and Hominids." *Scientific American* Vol. 250 (March 1984): 84–96.

Toon, Owen B. and Steve Olson. "The Warm Earth." *Science 85* Vol. 6 (October 1985): 50–57.

Towe, Kenneth M. "The Earth's Early Atmosphere." *Science* Vol. 235 (January 23, 1987): 415.

THE CRITICAL CYCLES

Berner, Robert A. and Antonio C. Lasaga. "Modeling the Geochemical Carbon Cycle." *Scientific American* Vol. 260 (March 1989): 74–81.

Cloud, Preston. "The Biosphere." *Scientific American* Vol. 249 (September 1983): 176–189.

Gale, George. "The Anthropic Principle." *Scientific American* Vol. 245 (December 1981): 154–171.

Kerr, Richard A. "No Longer Willful, Gaia Becomes Respectable." *Science* Vol. 240 (April 22, 1988), 393–395.

Monastersky, Richard. "The Whole-Earth Syndrome." *Science News* Vol. 133 (June 11, 1988): 393–395.

Safrany, David R. "Nitrogen Fixation." *Scientific American* Vol. 231 (October 1974): 64–80.

Schneider, Stephen H. "Climate Modeling." *Scientific American* Vol. 256 (May 1987): 72–80.

DANGER IN THE AIR

Flamsteed, Sam, "Ozone Ups and Downs." *Discover* Vol. 11 (January 1990): 33.

Graedel, Thomas E. and Paul J. Crutzen. "The Changing Atmosphere." *Scientific American* Vol. 261 (September 1989): 58–68.

Likens, Gene E., el al. "Acid Rain." *Scientific American* Vol. 241 (October 1979): 43–51.

Mohnen, Volker A. "The Challenge of Acid Rain." *Scientific American* Vol. 259 (August 1988): 30–38.

Monastersky, Richard. "Aerosols: Critical Question for Climate." *Science News* Vol. 138 (August 25, 1990): 118.

Raloff, Janet. "Mercurical Risks From Acid's Reign." *Science News* Vol. 139 (March 9, 1991): 152–156.

Roberts, Leslie. "Learning From an Acid Rain Program." *Science* Vol. 251 (March 15, 1991): 1302–1305.

Stolarski, Richard S. "The Antarctic Ozone Hole." *Scientific American* Vol. 258 (January 1988): 30–36.

MUDDYING THE WATERS

Bascom, Willard. "The Disposal of Waste in the Ocean." *Scientific American* Vol. 231 (August 1974): 16–25.

Editors. "Are Oil Spills an Environmental Hazard?" *Consumers' Research* Vol. 74 (January 1, 1991): 14–17.

la Riviere, J. W. Maurits. "Threats to the World's Water." *Scientific American* Vol. 261 (September 1989): 80–94.

MacIntyre, Ferren. "The Top Millimeter of the Ocean." *Scientific American* Vol. 230 (May 1974): 62–77.

Maranto, Gina. "The Creeping Poison Underground." *Discover* Vol. 6 (February 1985): 75–78.

_____. "A Once and Future Desert." *Discover* Vol. 6 (June 1985): 32–39.

Sun, Marjorie. "Ground Water Ills: Many Diagnoses, Few Remedies." *Science* Vol. 232 (June 20, 1986): 1490–1493.

Waters, Tom. "Ecoglasnost." *Discover* Vol. 11 (April 1990): 51–53.

THE DUMP SITE DILEMMA

Crawford, Mark. "Hazardous Waste: Where to Put It?" *Science* Vol. 235 (January 9, 1987): 156–157.

Grossman, Dan and Seth Hulman. "A Nuclear Dump: The Experiment Begins." *Discover* Vol. 10 (March 1989): 49–56.

———. "Down In The Dumps." *Discover* Vol. 11 (April 1990): 37–41.

Kao, Timothy W. and Joseph M. Bishop. "Coastal Ocean Toxic Waste Pollution: Where Are We and Where Do We Go?" *USA Today* Vol. 114 (July 1985): 20–23.

Marshall, Elliot. "The Geopolitics of Nuclear Waste." *Science* Vol. 251 (February 22, 1991): 864–867.

O'Leary, Philip R., Patrick W. Walsh, and Robert K. Ham. "Managing Solid Waste." *Scientific American* Vol. 259 (December 1988): 36–42.

Pfeiffer, Beth. "Our Disposable Society." *American Legion Magazine* (January 1990): 24–25 and 58.

Stone, Judith. "Toxic Avengers." *Discover* Vol. 10 (August 1989): 41–47.

DWINDLING RESOURCES

Davis, Ged R. "Energy for Planet Earth." *Scientific American* Vol. 263 (September 1990): 55–62.

Flower, Andrew R. "World Oil Production." *Scientific American* Vol. 238 (March 1978): 42–48.

Gibbons, John H., Peter D. Blair, and Holly L. Gwin. "Strategies for Energy Use." *Scientific American* Vol. 261 (September 1989): 136–143.

Griffith, Edward D. and Alan W. Clarke. "World Coal Production." *Scientific American* Vol. 240 (January 1979): 38–47.

Marlay, Robert C. "Trends in Industrial Use of Energy." *Science* Vol. 226 (December 14, 1984): 1277–1282.

Hubbard, Harold M. "The Real Cost of Energy." *Scientific American* Vol. 264 (April 1991): 36–42.

Rosenfield, Arthur H. and David Hafemeister. "Energy-efficient Buildings." *Scientific American* Vol. 258 (April 1988): 78–85.

Woche, Wirtschaffs. "Sea Riches: What Future?" *World Press Review* Vol. 31 (November 1984): 23–25.

BURGEONING POPULATIONS

Berreby, David. "The Numbers Game." *Discover* Vol. 11 (April 1990): 43–49.

Brown, Lester R. "World Population Growth, Soil Erosion, and Food Security." *Science* Vol. 214 (November 27, 1981): 995–1001.

Coale, Ansley J. "The History of the Human Population." *Scientific American* Vol. 231 (September 1974): 41–51.

Crosson, Pierre R. and Norman J. Rosenberg. "Strategies for Agriculture." *Scientific American* Vol. 261 (September 1989): 128–135.

Demeny, Paul. "The Populations of the Undeveloped Countries." *Scientific American* Vol. 231 (September 1974): 149–159.

Gibbons, Boyd. "Do We Treat Our Soil Like Dirt." *National Geographic* Vol. 166 (September 1984): 353–388.

Gwatkin, Davidson R. and Sarah K. Brandel. "Life Expectancy and Population Growth in the Third World." *Scientific American* Vol. 246 (May 1982): 57–65.

Keyfitz, Nathan. "The Growing Human Population." *Scientific American* Vol. 261 (September 1989): 119–135.

Revelle, Roger. "Food and Population." *Scientific American* Vol. 231 (September 1974): 161–170.

THE LOSS OF LIFE

Barinaga, Marcia. "Where Have All the Froggies Gone?" *Science* Vol. 247 (March 2, 1990): 1033–1034.

Cohn, Jeffrey P. "Gauging the Biological Impacts of the Greenhouse Effect." *BioScience* Vol. 39 (March 1989): 142–146.

Colinvaux, Paul A. "The Past and Future Amazon." *Scientific American* Vol. 260 (May 1989): 102–108.

Diamond, Jared. "Playing Dice with Megadeath." *Discover* Vol. 11 (April 1990): 55–59.

Kunzig, Robert. "Invisible Garden." *Discover* Vol. 11 (April 1990): 67–74.

Monastersky, Richard. "Swamped by Climate Change?" *Science News* Vol. 138 (September, 22, 1990): 184–186.

Repetto, Robert. "Deforestation in the Tropics." *Scientific American* Vol. 262 (April 1990): 36–42.

Roberts, Leslie. "Is There Life After Climate Change?" *Science* Vol. 242 (November 18, 1988): 1010–1013.

Wilson, Edward O. "Threats to Biodiversity." *Scientific American* Vol. 261 (September 1989): 108–116.

HOT AS A GREENHOUSE

Ellsaesser, Hugh W. "The Greenhouse Effect: Science Fiction?" *Consumers' Research* Vol. 71 (November 1988): 27–31.

Jones, Philip D. and Tom M. L. Wigley. "Global Warming Trends." *Scientific American* Vol. 263 (August 1990): 84–91.

Ramanathan, V. "The Greenhouse Theory of Climate Change: A Test by Inadvertent Global Experiment." *Science* Vol. 240 (April 15, 1988): 293–299.

Revelle, Roger. "Carbon Dioxide and World Climate." *Scientific American* Vol. 247 (August 1982): 35–43.

Revkin, Andrew C. "Endless Summer: Living with the Greenhouse Effect." *Discover* Vol. 9 (October 1988): 50–61.

Schneider, Stephen H. "The Greenhouse Effect: Science and Policy." *Science* Vol. 243 (February 1989): 771–779.

_____. "The Changing Climate." *Scientific American* Vol. 261 (September 1989): 70–79.

Tangley, Laura. "Preparing for Climate Change." *BioScience* Vol. 38 (January 1988): 14–18.

White, Robert. "The Great Climate Debate." *Scientific American* Vol. 263 (July 1990): 36–43.

THE SATELLITE SOLUTION

Booth, William. "Monitoring the Fate of the Forests from Space." *Science* Vol. 243 (March 17, 1989): 1428–1429.

Hardisky, M. A., M. F. Gross, and V. Klemas. "Remote Sensing of Coastal Wetlands." *BioScience* Vol. 36 (July/August 1986): 453–458.

Paul, Charles K. and Adolfo C. Mascarenhas. "Remote Sensing in Development." *Science* Vol. 214 (October 1981): 139–145.

Perry, Mary Jane. "Assessing Marine Primary Production from Space." *BioScience* Vol. 36 (July/August 1986): 461–466.

Rock, B. N., et al. "Remote Detection of Forest Damage." *BioScience* Vol. 36 (July/August 1986): 439–444.

Roller, Norman E. G. and John E. Colwell. "Coarse-Resolution Satellite Data for Ecological Surveys." *BioScience* Vol. 36 (July/August 1986): 468–472.

H. Yates, el al. "Terrestrial Observation from NOAA Operational Satellites." *Science* Vol. 231 (January 31, 1986): 463–469.

CLEANING UP OUR ACT

Booth, William. "Johnny Appleseed and the Greenhouse." *Science* Vol. 242 (October 7, 1988): 19–20.

Clark, William C. "Managing Planet Earth." *Scientific American* Vol. 261 (September 1989): 47–54.

Dickson, David. "Europe Struggles to Control Pollution." *Science* Vol. 234 (December 12, 1986): 1315–1316.

Marshall, Elliot. "EPA's Plan for Cooling the Global Greenhouse." *Science* Vol. 243 (March 24, 1989): 1544–1545.

Raloff, Janet. "Urban Smog Control: A New Role for Trees." *Science News* Vol. 138 (July 7, 1990): 5.

Ruckelshaus, William D. "Toward a Sustainable World." *Scientific American* Vol. 261 (September 1989): 166–174.

Weisskopf, Michael. "Cooling Off the Greenhouse." *Discover* Vol. 10 (January 1989): 30–33.

GLOBAL POLLUTION

Elmer-Dewitt, Philip. "A Man-Made Hell on Earth." *Time* Vol. 137 (March 18, 1991): 36–37.

Ehrlich, Paul R., et al. "Long-Term Biological Consequences of Nuclear War." *Science* Vol. 222 (December 23, 1983): 1293–1299.

Lacayo, Richard. "A War Against the Earth." *Time* Vol. 137 (February 4, 1991): 32–33.

Monastersky, Richard. "Global Change: The Scientific Challenge." *Science News* Vol. 135 (April 15, 1989): 232–235.

Raloff, Janet and Richard Monastersky. "Gulf Oil Threatens Ecology, Maybe Climate." *Science News* Vol. 139 (February 2, 1991): 71–73.

Schell, Jonathan. "Our Fragile Earth." *Discover* Vol. 10 (October 1989): 45–50.

Turco, David P. "The Climatic Effects of Nuclear War." *Scientific American* Vol. 251 (August 1984): 33–43.

Usher, Peter. "World Conference on the Changing Atmosphere: Implications for Global Security." *Environment* Vol. 31 (January/February 1989): 25–27.

Index